Class	MAMMALIA	
Subclass	PROTHERIA	
Order	†Triconodonta	
	†Docodonta	
	Monotremata	(platypus, spiny anteaters)
	†Multituberculata	
Subclass	THERIA	
Infraclass	†Pantotheria	
	Marsupialia	(opossums, kangaroos etc)
	Eutheria	
Cohort	EDENTATA	
Order	Xenarthra	(anteaters, sloths and armadillos)
Cohort	EPITHERIA	
Grandorder	†ICTOPSIA	
	ANAGALIDA	
Order	Macroscelidea	(elephant-shrews)
	Lagomorpha	(rabbits, hares and pikas)
Grandorder	FERAE	
Order	†Cimolesta	
	†Creodonta	
	Carnivora	(dogs, weasels, cats, seals etc)
Grandorder	INSECTIVORA	(hedgehogs, moles, shrews etc)
	ARCHONTA	
Order	Scandentia	(tree shrews)
	Dermoptera	(flying lemurs)
	Chiroptera	(bats)
	Primates	(man, apes, monkeys, lemurs etc)

Grandorder	UNGULATA	
Order	†Taeniodonta	
	†Pantodonta	
	†Arctocyonia	
	†Tillo	
	Tub	
	†Din	
	†Emb	
	Artio	
	†Acreo	
	Cetacea	(whales, dolphins, porpoises etc)
	†Litopterna	
	†Notoungulata	
	†Astrapotheria	
	†Trigonostylopoidea	
	†Xenungulata	
	†Pyrotheria	
	†Condylarthra	
	Perissodactyla	(odd-toed ungulates: horses, tapirs, rhinos etc)
	Hyracoidea	(hyraxes)
	Proboscidea	(elephants)
	Sirenia	(sea cows)
	†Desmostylia	
Grandorder	incertae sedis	
Order	Pholidota	(scaly anteaters)
	Rodentia	(squirrels, rats, mice etc)

† extinct

Mammal evolution
an illustrated guide

Text by R.J.G. Savage

Professor of Geology, University of Bristol

&

Illustrations by M. R. Long

Senior Lecturer, Brunel College, Bristol

Facts On File Publications
New York, New York ● Oxford, England

Mammal Evolution: An Illustrated Guide

A copublication of Facts On File and
The British Museum (Natural History)

Library of Congress Cataloging-in-Publication Data

Savage, Robert J. G.
  Mammal evolution.

  Bibliography: p. 251
  Includes index.
1. Mammals–Evolution.   2. Mammals, Fossil.
I. Long, M.R.   II. title.
QL708.5.S28   1986   599'.03'8   85-29203
ISBN 0-8160-1194-X (Facts on File)

Printed in the United Kingdom
10 9 8 7 6 5 4 3 2 1

Contents

Preface

...and over these
The jagged alligator, and the might
Of earth-convulsing behemoth, which once
Were monarch beasts, and on the slimy shores,
And weed-overgrown continents of earth,
Increased and multiplied like summer worms
On an abandoned corpse, till the blue globe
Wrapped deluge round it like a cloak, and they
Yelled, gasped, and were abolished; or some God
Whose throne was in a comet, passed, and cried,
'Be not!' And like my words they were no more.

PROMETHEUS UNBOUND
Percy Bysshe Shelley 1820

Shelley's lines paint a picture of a lost world, a world resurrected from the antiquities of Greek mythology. But Shelley had more than ancient legends to go on; he was acutely aware of the discoveries of great extinct beasts from the writings of contemporary naturalists. It is known that Shelley attended lectures on mineralogy at Oxford and possessed a library of scientific works which included the writings of Newton, Laplace, Herschel, Davy and Erasmus Darwin. By 1812 he was familiar with James Parkinson's *Organic remains* published in three volumes between 1804 and 1811. Parkinson was a surgeon and an oryctologist (one who studies fossils or more literally 'things dug up'; the term was popular in the early nineteenth century to be ousted soon after by the term palaeontology). In *Organic remains* we find vivid portraits of strange gigantic beasts destroyed by a great deluge, the visible evidence of the biblical flood. Our aim here is also to record lost worlds, but in prose rather than poetry, and in anatomical detail rather than in imaginative verse. Shelley's works, however, demonstrate all too clearly how in the early nineteenth century new scientific knowledge was eagerly acquired by the *cognoscenti* when there were no fissures separating arts and sciences.

To give the kiss of life to a fossil may seem a futile exercise, but patient study of those old bones enables us to clothe them in flesh and see how they performed in life. Our story is the life history of mammals. Some 220 million years ago mammals appeared on the face of the earth; they like mice kept a very low profile for the next 150 million years while dinosaurs reigned supreme as the lords of creation. But as the dinosaurs took their final curtain call, the mammals waiting in the wings appeared on the stage to replace them, conquering all lands and invading seas and skies. It is the story of that great series of radiations – of bats and whales, of mice and men, of elephants and tigers, of camels and kangaroos, of mastodonts and glyptodonts – that we recount. It is a long, exciting and vastly complex story.

The first four chapters deal with the essential background; how mammals become fossilized, how mammals compare with other animals, how we re-clothe fossil bones with flesh and how a group of reptiles evolved into the first mammals. The grouping of mammals that follows is based primarily on how the animals feed. We look as those that eat insects and flesh, those that gnaw, those vegetarians that feed on roots and tubers, the browsers and the grazers. Separately we consider fliers and swimmers and finally monkeys and men. This last chapter is brief because of all mammal stocks, this is the one that has a voluminous popular literature.

Throughout the book there is a two-fold aim: to give a readily digestible survey of the subject at a level that will we hope inform all but the specialists, and to illustrate the examples lavishly with restorations, with skeletons and dentitions – for it is bones and teeth upon which almost all our evidence rests. These are supplemented as needs be by family trees, maps and diagrams. The book contains a glossary to help reading for the novice and a comprehensive list of references is supplied for further reading.

Writing and illustrating the book has been fun. Preparation for publication has involved intriguing callisthenics. Colour illustrations appear on every second fold and allying the text and illustrations to this format has not always been easy. As a convention dentitions are almost always illustrated pointing with the front teeth to the right side of the page; the upper dentitions are drawn as from the right side and the lower dentitions as from the left side – in this way the two halves can be superposed and the cusp relationships more easily interpreted. All restorations throughout the book are original and all figures are original or redrawn; in the latter case acknowledgement is given to the source of the original.

Book production relies heavily on team work and we have been extremely fortunate in having unfailing

courtesy, enthusiastic support and patient
understanding from the publications staff at the British
Museum (Natural History). In thanking them all
warmly we would especially like to mention Dr Gordon
Corbet, Mr Robert Cross, Mr Chris Owen and Mr Eric
Dent. We also acknowledge with thanks the
contributions to the illustrations from Alma Gregory,
Graham Flowerdew, Paul Weaver, Timothy Duke,
Simon Long and James Long.

We are grateful to the relevant scholars at the British
Museum (Natural History) for their comments on drafts
of the text and illustrations. We the authors are,
however, solely responsible for all remaining errors.

CHAPTER 1
Bones into stones

On the East African plains a lioness makes a kill; the family feed on the zebra while hyaenas and jackals lie in wait, to be followed by vultures and lesser predators. After a few hours the dismembered carcass is stripped of its meat, left to the beetles, and to the elements – sun, wind and rain. It is a scenario that is repeated thousands of times every day in all parts of the world with carnivores of all sizes from tigers to weasels. Survival for some means death for others. Death provides the raw material for fossilization, but the chances of a complete mammal becoming fossilized are very low; even for a few teeth to survive for posterity is a rare event. Death is always accompanied by the onset of destructive events, biological and physical.

Forms of preservation

The two most important factors operating in favour of bone preservation are water transport and sediment cover. Without these, mammals and other animals dying on land will be scavenged and the remains rot away under the elements. But if transported while still fresh into a lake or lagoon, their remains begin to fossilize. Most mammals live on land and their remains are fossilized in freshwater deposits, though occasionally they may be carried by rivers to the sea. They rarely survive long enough to fossilize in the open sea, but if the river issues into a lagoon their chances of burial are good. Once buried they quickly lose any remaining organic matter and the calcium phosphate of their bone often becomes impregnated with minerals, especially iron; hence the black colouring of many fossil bones.

An unusual form of preservation is found in the Rancho la Brea tar pits in California, where mammals coming down to pools to water got stuck in the gummy tar and drowned; carnivores and vultures after easy prey also got caught. The asphalt is an extremely good fossilizing element and remains from the pits are very abundant; whole skeletons can be reconstructed and among these sabretooth cats are the most spectacular.

Clay muds compact into mudstones and shales.

Lime muds compact into limestones.

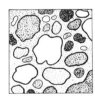

Quartz grains form sands which on compaction become sandstones.

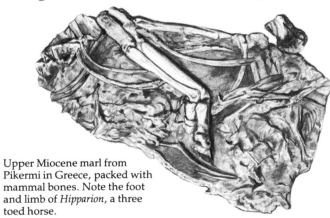

Upper Miocene marl from Pikermi in Greece, packed with mammal bones. Note the foot and limb of *Hipparion*, a three toed horse.

Most sediments are derived from three distinct materials, either singly or in combination. In the course of time these become compacted resulting in the reduction of pore spaces and in their cementation to form rocks.

Baby mammoth from Siberia, found in 1977. It had fallen into a crevasse thousands of years ago and was deep frozen in the ice.

Mammoth from Berezovka in NE Siberia; it broke a leg and hipbone in falling into the crevasse, and the trunk was eaten by wolves before scientists recovered it for the Leningrad Museum in 1901.

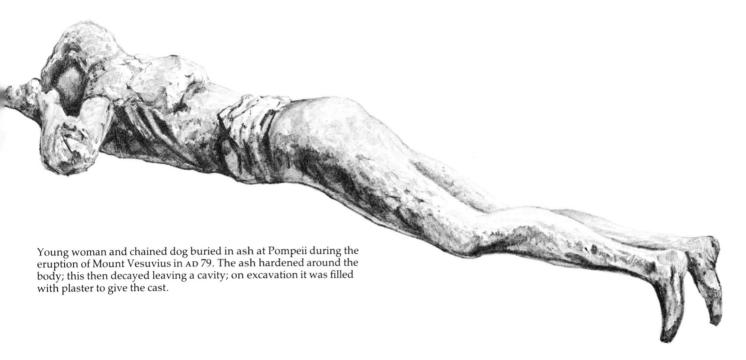

Young woman and chained dog buried in ash at Pompeii during the eruption of Mount Vesuvius in AD 79. The ash hardened around the body; this then decayed leaving a cavity; on excavation it was filled with plaster to give the cast.

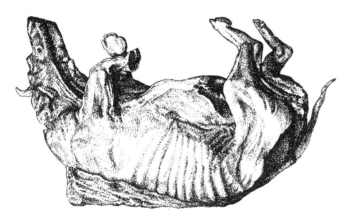

Woolly rhinoceros: a cadaver from a Pleistocene site at Starunia in Galicia, Poland. The cadaver is preserved by impregnation with salt and petroleum.

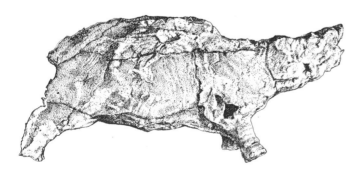

As the early Miocene lavas advanced near Blue Lake, Washington, USA, they entombed a shallow pond in which the dead rhinoceros was floating; the water cooled the lavas which formed pillows; these moulded themselves around the carcass, thus preserving it in death pose.

Sands and lime muds, becoming sandstones and limestones, are the commonest media for preservation. The cementing material is usually calcite (calcium carbonate). Fortunately calcite is soluble in organic acids and bone (calcium phosphate) is not, so playing on this chemical difference, we can extract whole skeletons of even the most fragile mammals.

Another common component in fossiliferous sediments is volcanic ash. Many of the famous Tertiary (65–1·9 million years ago) mammal sites in East Africa and the Rockies of North America are of ash deposited in lakes; the ash from volcanoes providing rapid burial in very fine grained material which preserves all the bone detail. At Olduvai Gorge in Tanzania thousands of mammals are thus preserved in deposits one to three million years old and with them are found remains of early man. Another famous ash site is at Herculaneum near Pompeii in Italy, where the eruption of AD 79 buried people and their possessions. The people (and their dogs) survived long enough for the ash to harden around them, so that when it is removed, the vacant space found is a replica. This occasionally happens with lava, as with the rhinoceros discovered in a lava flow in the State of Washington, USA.

During the ice ages of the last two million years mammoths walking across a glacier occasionally fell into crevasses, became entombed and deep frozen. There they remained for thousands of years, only to be uncovered when the ice melted or a river cut a new course through the permafrost, as is the case with some Siberian and Alaskan mammoths. This preservation of a complete carcass allows a detailed study of the soft anatomy to be made, even to a study of the hair, and stomach contents; in summer, mammoths fed on grassy meadows.

Also during the ice ages, many potholes exposed on the limestone plateaux acted as traps and mammals occasionally fell down into the caves below. Bears and other mammals may shelter in the cave entrances or

Scene from Les Combarelles in the Dordogne, France with mammoth and two horses, one horse curiously has a bovid horn. Engraved and black paint. Upper Palaeolithic age.

Skin of extinct ground sloth from a cave in Patagonia, South America. The skin has a thick layer of coarse hair and imbedded within the skin are nodules of bone.

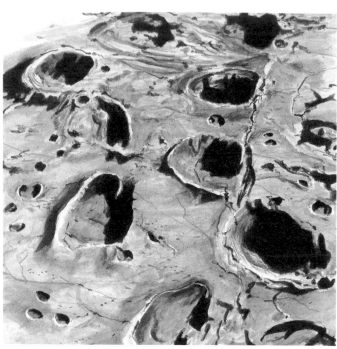

Footprints of giraffe and small antelope from Laetoli, Tanzania, made some three million years ago as they walked over damp soil composed largely of volcanic ash.

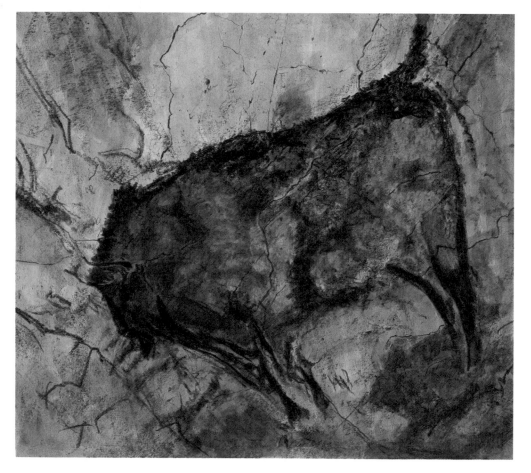

Polychrome painting of a bison from Altamira Cave in north Spain. These paintings date from the Magdalanian period some 15 000 years ago and represent the most advanced form of cave art.

Reconstruction of a scene at a tar pool at Rancho La Brea. The mammoth has become trapped in the oily waters and is being devoured by the sabretooth cat *Smilodon*, while the vultures and dire wolves wait their turn to feed on the carcass.

even hibernate there. Over thousands of years the remains of those that fail to survive the winter accumulate; the bone caves of the Mendip Hills and of Torquay in England, and the Drachenhöle in Austria are examples. Caves in Patagonia, South America have yielded skin of extinct giant ground sloths – preserved under very dry desert conditions; natural mummification.

Indirect evidence of mammals is found in footprints, preserved when muds over which the mammals walked, harden quickly and are then covered by soft sediments before they can be destroyed. At Laetoli in Tanzania, Dr Mary Leakey has recently described a fauna of savannah mammals some 3·5 million years old from extensive pavements of footprints.

Palaeolithic man depicted his game on the walls of caves; from France and Spain we have superb polychrome picture galleries of mammoths, reindeer, woolly rhinoceroses and many other mammals, many species of which are now extinct.

The restless earth

Agents of weathering are continuously at work – wind, frost, sun, rain, rivers and so on. Their action is physically and chemically to reduce the rocks and transport them into the seas. If the processes were to continue uninterrupted the continents would soon be reduced to lowland plains and the seas choked with sediments. But forces within the earth are constantly countering these activities. The crust of the earth can be compared with the skin of a desiccated apple; in drying the skin cracks and becomes wrinkled. So with the earth's crust which comprises a number of segments or plates. The mantle below the crust is hot and mobile, and slowly moves the crustal fragments around.

There are two sorts of crust, continental underlying the continents and oceanic underlying the oceans. The plate margins may be in mid ocean, mid continent or at the ocean/continent junction. The movements of the plate margins can be divergent, convergent or they can shear past each other. All these are illustrated in the accompanying figures. Active movement at plate margins gives rise to two spectacular phenomena – earthquakes and volcanoes. Earthquakes occur when

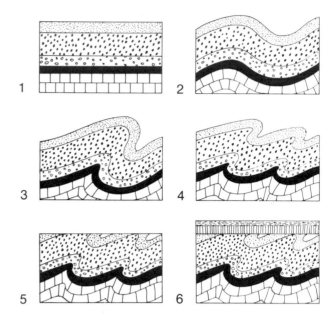

Sediments are laid down horizontally (1). In active regions of the crust they become deformed, that is folded, compressed and even fractured, as a table cloth does when its sides are pushed towards the centre of the table. At first the folds are gentle (2), then steepen (3) and later become overturned (4). After a lapse of time these folded rocks may be exposed to weathering and eroded (5); then a second sequence of sediments may be laid down unconformably on top (6).

great stress is suddenly released and a geological fault is the result; such is the San Andreas fault in California. Active volcanoes with lava flows (as Mt Heckla in Iceland) and ash falls (as Mt St Helens in the State of Washington) are produced when mantle magmas rise violently to the surface.

In addition to the extrusion of lavas which form great basalt plateaux, other molten rocks may solidify beneath the surface of the crust and later be exposed on weathering as granites. Rocks sucked down deep into the bowels of the earth on plate collision, become metamorphosed by heat and pressure, and they too may later surface during mountain building processes as gneiss and schists, such as cover much of Scotland, Norway and Canada. These metamorphic and igneous rocks do not normally preserve fossils, so it is almost exclusively in the sedimentary rocks, the products of weathering, that we must search for fossils.

Stair Hole at Lulworth in Dorset, England. This is the Lulworth Crumple, an example of complex folding and fracturing of Purbeck (Jurassic) rocks. The beds were folded in Miocene times while the Alps were being elevated in southern Europe.

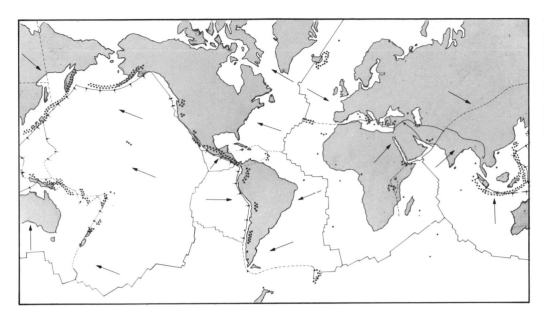

The crust of the earth is composed of seven major lithospheric plates, bounded by active ridge crests, transform faults, trench systems and zones of compression. In addition there are five minor plates. The directions of movement are indicated by arrows; the annual rates of movement vary from 1 cm (e.g. Iceland) to about 10 cm (e.g. equatorial Pacific).

Most sediments are carried by rivers into the oceans and the remains of land-dwelling animals rarely survive the long journey; hence mammals in marine deposits are rare, save for occasional whales and other marine mammals. However, in areas where rivers issue into lagoons, the consequent energy reduction causes their load to be dropped. After predation, the second commonest cause of death in mammals is drowning; in savannah and semi-desert terrains the vegetation is often richest in the dry river beds. Sudden storms bring down a spate of water that sweeps the browsing and grazing animals off their feet; they drown and are carried for long distances downstream. Should the river empty into a lake this can provide an ideal environment for preservation. Should there also be volcanoes showering out ash that gets washed into the waters of the lake so much the better, for it provides rapid burial and hence good preservation. Such lacustrine ash deposits account for many of the great successions of Tertiary mammal faunas in the Rockies of the western USA, in the Pampas of Argentina and in the Rift Valley of East Africa.

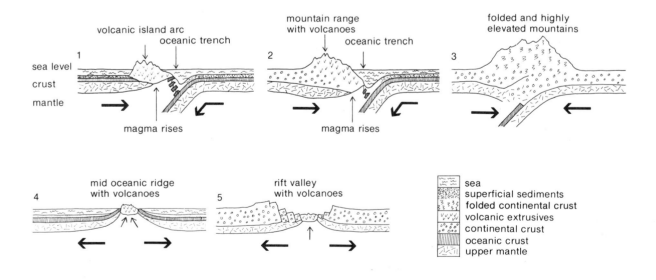

The five major types of plate junction illustrated diagrammatically.
1 Convergent oceanic crusts, producing island volcanic arcs with subduction beneath as witnessed by a nearby deep sea trench, e.g. Indonesia, south-east Asia.
2 Convergent continental and oceanic crusts, producing mountain ranges with volcanoes and off-shore deep sea trench at subduction junction, e.g. Pacific coast of South America.
3 Convergent continental crusts, producing very high folded mountain ranges, e.g. Himalayas as India and Asia converge.
4 Divergent oceanic crust, producing mid-oceanic volcanic ridge. e.g. Atlantic Ocean with its submarine ridge.
5 Divergent continental crust, producing rift valleys. e.g. East African Rift Valley.

Reconstruction of a scene in Antelope County, Nebraska some 10 million years ago as a herd of rhinoceroses (*Teleoceras*) were overcome with a volcanic ash fall.

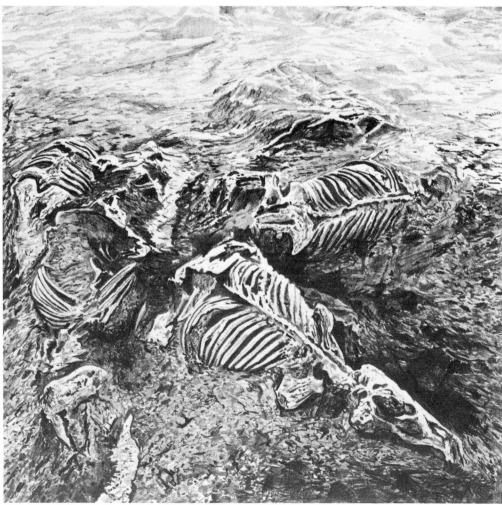

Nebraska in the late Miocene some 10 million years ago when thick falls of volcanic ash killed a large population of rhinoceroses, horses and camels.

Dating the rocks

The scientific study of geological history goes back almost 200 years, and we owe our basic understanding of the succession of sedimentary rocks to William Smith, a canal engineer. Around the 1790s he was working in the Bath area and from his experience of the rocks there he noted similarities to these and their fossils in other places as he travelled widely across England. This led him to formulate the two basic laws of stratigraphy: (1) that strata are superimposed, the younger beds being on top; (2) that each stratum has its own characteristic assemblage of fossils, which enables it to be correlated with strata of similar age in other places. These laws were the foundation for the drawing up of a table of strata which Smith made in 1799. This table or stratigraphic column in its modern form divides geological time into a series of eras and periods. From the Pre-Cambrian eras we have little evidence of organic life. During the Palaeozoic era the dominant vertebrates were the fishes, with amphibians from Carboniferous times onward. During the Mesozoic era the reptiles were the ruling vertebrates on land with mammals present but rarely bigger than a rat in size and never very abundant. It was during the Caenozoic era that the mammals became the dominant land vertebrates, with birds, insects and flowering plants playing important roles in the ecology of their life cycles.

During the nineteenth century it was possible thus to correlate strata, and thence their fossils, relative to each other, but there was no way of estimating how old they were. Was a dinosaur one thousand, one million or one hundred million years old? With the discovery of radioactivity early in this century it became possible to date rocks absolutely, since radioactive minerals decay at known rates into stable end products. By measuring the proportions of each, the time elapsed since the mineral was formed can be calculated. The radioactive uranium, thorium and potassium minerals are usually found in igneous rocks which do not contain fossils; however lavas between the sediments can be dated and volcanic ash deposits with fossils can themselves be dated. The Caenozoic era is known to be some 65 million years old, though the earliest mammals are found in rocks some 220 million years old. The enormous expanses of time are difficult if not impossible to comprehend, so let us put them on a scale. If we let the whole of geological time since the origin of the earth around 4500 million years ago be represented by one calendar year, we have a time scale of one day representing 12·3 million years. On this scale with the origin of the earth on 1 January, the earliest Cambrian fossils appeared on 13 November. The first mammals appeared on 14 December, the Caenozoic began on 25 December, and the first man appeared at 5 p.m. on 31 December.

The Geological Column. The scale is logarithmic, beginning from the present, so that the earlier and longer periods occupy less space.

Geological Column					
Eras		Periods	Million Years (MA)	Calendar Year	hours
Caenozoic	Quaternary	Holocene	0.01	31 December	23.59
		Pleistocene	1.9	31 December	20.18
	Tertiary	man			
		Pliocene	5.1	31 December	14.38
		hominids Miocene	24	29 December	
		Oligocene	37	28 December	
		Eocene	58	26 December	
		primates Palaeocene	65	25 December	
Mesozoic					
		Cretaceous	145	19 December	
		Jurassic	215	13 December	
		mammals Triassic	250	10 December	
Palaeozoic		Permian	285		
		reptiles Carboniferous	360		
		amphibians Devonian	410	28 November	
		jawless fishes Silurian	440		
		Ordovician	505		
		Cambrian	590	13 November	
Precambrian					
		microorganisms First fossil	3500	22 March	
		Oldest rock	4000	9 February	
		Earth's origin	4500	1 January	

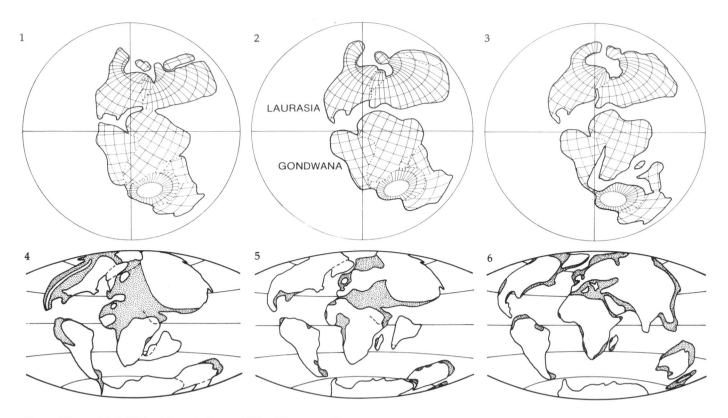

Maps of the world at different times in the past 175 million years. The shaded areas indicate shallow epicontinental seas and the dotted lines indicate present-day coastlines. (4–6 after Cox, Healey & Moore).

1 Middle Jurassic 175 MYA
  Pangaea is fragmenting with seas encroaching between the two Americas

4 Late Cretaceous 75 MYA
  Present-day Africa has rejoined with Asia and the south Atlantic Ocean has begun to open up

2 Early Cretaceous 135 MYA
  Pangaea has fragmented into the northern Laurasia and the southern Gondwana super-continents

5 Middle Eocene 50 MYA
  The Indian subcontinent has parted company with Africa and is drifting northwards. Australia has separated from Antarctica

3 Middle Cretaceous 100 MYA
  Laurasia is beginning to break up into North America and Eurasia. Gondwana has deep fractures and the continental blocks are separating

6 Early Miocene 20 MYA
  The present-day distribution of the continents is clearly recognizable. In its union with Asia, India has thrown up the Himalayas.

Mammals around the world

Mammals are known on all continents and are only absent from oceanic islands. Antarctica has today no true land mammals, but abounds in seals. Land mammals are known in Eocene sediments on Antarctica.

The most important geological event in the early history of mammals was the break up of the supercontinent Pangaea. During late Carboniferous and Permian times there was one great supercontinent Pangaea and one great super-ocean Panthalassa. This situation persisted through most of Triassic times. Towards the end of Triassic and in early Jurassic times cracks appeared in the continental crust; rift faulting with associated vulcanism developed in the southern North Atlantic area. In New Jersey the Newark basin developed and was infilled during the mid and late Triassic with river sediments. In the early Jurassic these sediments were faulted and the great 300-metre Palisade igneous sill intruded.

The earliest known mammals date from this late Triassic–early Jurassic period and they are known from North America, Europe, eastern Asia and South Africa. Their ancestral stocks, the cynodont reptiles, are also recorded from the same areas and additionally from South America. Together these facts establish that the mammals had become effectively worldwide in distribution before Pangaea broke up.

By mid Jurassic times the seas were well established in the area between eastern USA and West Africa as the two continents separated. During the later Mesozoic the North Atlantic opened up as Europe and North America separated. The break up of the southern landmass – Gondwana – is complex and more difficult to date. The presence of marine faunas of early Jurassic age in Kenya and Tanzania implies at least partial separation of Africa and India. The opening of the South Atlantic can be traced through Cretaceous times. The isolation of Antarctica is probably as late as Oligocene. As our knowledge grows of the geology of the ocean floors, so will we be better able to reconstruct palaeogeographic maps. As yet the geological history and the mammal distributional history are often too poorly known to reconstruct a clear and integrated sequence of events.

CHAPTER 2
Mammals as animals

Naming mammals

Several million kinds of plants and animals are known today, and there are many more millions that are extinct. Man has named and attempted to classify them since ancient times. We find accounts of the different kinds of animals around the world in the Upanishads, the Bible, the Koran, in the writings of Aristotle and Pliny. However, from a scientific as opposed to a historic viewpoint, we begin with the work of the Swedish naturalist Carl Linnaeus (1707–1778). In order to recognize and refer to animals and plants we give them names, just as we give people names; this is what Linnaeus did in his *Systema Naturae* and to this day the twelfth edition (1758) of his great work is still taken as the base line for zoological nomenclature.

Linnaeus did three things in that work: he gave every known kind of organism, living and fossil, a binomial Latin name; he described its characteristics, differentiating it from other organisms similar to it; and he placed all organisms in a hierarchical classification.

Each species has two names – the generic and the trivial. For example the dog is *Canis familiaris*. *Canis* is the genus or generic name; it is always written with a capital letter. The trivial name *familiaris* is never written with a capital letter, and together they make up the species, which is distinct from *Canis lupus* the wolf or *Canis mesomelas* the jackal. These three are all closely related; they share the generic name, like John, William and Harry Smith. It is practice throughout the scientific world to refer to organisms by their binomial Latinized name – then no misunderstandings can arise from difficulties of translation between languages, where

often a native name does not exist – there is no Eskimo word for an elephant or giraffe, and in Swahili *mbwa* may mean any sort of dog, fox or jackal. Generic names are unique and may be used on their own; *Canis* can thus refer to any of about 40 species included within the genus. However, the trivial name cannot be used in isolation for it is not unique. *Elephas maximus* is the Indian elephant while *Priodontes maximus* is the giant armadillo of South America.

Linnaeus differentiated his species on their different characteristics, mainly differences of form, appearance, behaviour, and geographic distribution. *Canis lupus* occurs in northern latitudes while *C. mesomelas* occurs in Africa. *Camelus dromedarius* has one hump while *C. bactrianus* has two humps.

The next thing is to group them in such a way that they are readily referable to and for this we use a hierarchical system of classification. So we can group all the species of *Canis*, then all the genera of canid (foxes, raccoon, dog, maned wolf, Cape hunting dog and so on) in the family Canidae. With the Canidae, we group other carnivorous families – Ursidae (bears), Felidae (cats), Hyaenidae (hyaenas), Viverridae (civets and mongooses), Procyonidae (raccoons), Mustelidae (stoats and otters), Phocidae (seals) and Otariidae (sealions) – all together in the order Carnivora. There are about 40 orders (e.g. elephants, rodents, cattle, primates), which comprise the class Mammalia. The mammals, together with fish, amphibia, reptiles and birds and make up the phylum Chordata or the vertebrates as they are known in the vernacular to distinguish them from all the other phyla of invertebrate animals (insects, molluscs etc.).

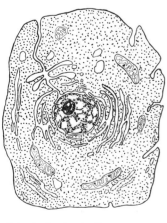

A typical animal cell as seen under a light microscope. The cell is bounded by a membrane and contains protoplasm, the two major components of which are cytoplasm and the nucleus. Other features depicted can be seen by using special staining techniques.

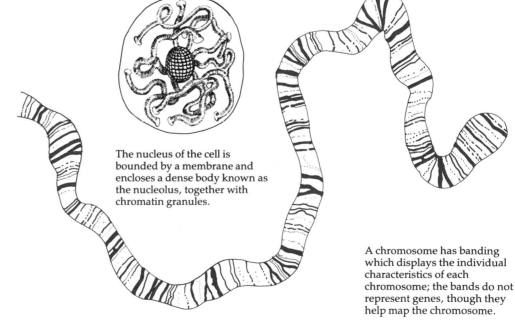

The nucleus of the cell is bounded by a membrane and encloses a dense body known as the nucleolus, together with chromatin granules.

A chromosome has banding which displays the individual characteristics of each chromosome; the bands do not represent genes, though they help map the chromosome.

The structure of a DNA molecule. The sides of the spiral ladder or double helix are made up of alternating sugar and phosphate pieces. The rungs or threads of the ladder are made of four kinds of bases which pair up such that adenine always links with thymine and cytosine with guanine.

Chromosomes from a male human cell, enlarged about 3000 times. There are 22 pairs of auto chromosomes and a pair of sex chromosomes (X and Y) making 46 in total.

Chromosomes from cell of a mouse; there are 38 auto chromosomes and a pair of sex chromosomes.

Classification of the dog

KINGDOM	Animalia
PHYLUM	Chordata
CLASS	Mammalia
ORDER	Carnivora
FAMILY	Canidae
GENUS	*Canis*
SPECIES	*Canis familiaris*

Classification is based on a study of the similarities and differences seen in each species and this Linnaeus believed reflected degrees of relationship. The more they have in common, the more closely related they will be; this gives us a natural or phylogenetic classification. But the significance of that relationship was not appreciated until Charles Darwin proposed the theory of evolution. Linnaeus had considered species as static, fixed, natural entities. Darwin saw species as dynamic entities, gradually changing one into another. It was often said that God created life, but Linnaeus created order.

Evolution and natural selection

Charles Darwin (1809–1882) published his famous book *On the origin of species* in 1859, almost exactly a century after Linnaeus's twelfth edition. In his book Darwin was concerned with the origin of species and proposed a mechanism – natural selection. The theory can be outlined in three statements based on observations and two deductions.

STATEMENT 1: in nature, the number of offspring is always greater than the parental number. For example, an oak tree produces many more than one acorn; a salmon lays many more than two eggs, a rabbit has many more than two young in a life-time.

STATEMENT 2: the number of individuals of a species remain more or less constant in time. Without the interference of man, the number of oak trees, salmon and rabbits does not change greatly over the years, though there may be highs if food supplies are good and lows if disease strikes.

DEDUCTION 1: from these observations Darwin deduced there was a struggle for survival.

STATEMENT 3: no two individuals are identical (other than identical twins); that is all organisms vary.

DEDUCTION 2: in the struggle for survival, some varieties are favoured, i.e. natural selection ensures the survival of the fittest.

In Darwin's day the mechanisms of inheritance were not understood. We would now recognize two sorts of variation, that due to environment and that due to inheritance, in other words nurture and nature. Old men are often portly and bald. Their portliness is due to nurture and their baldness to the nature of their genetic inheritance.

All organisms are built up of cells; the cells all have a nucleus; the nucleus contains chromatin, which, when the cell divides is organized into pairs of chromosomes, half the number (n) going to each sex cell, so that when the male sperm and female ovum fuse the full complement ($2n$) will again be present, with half deriving from each parent. The chromosomes are made up of genes; in man there are 44 chromosomes plus the sex determining pair ($2n = 46$); some mammals have as many as 84 (white rhinoceros), some as few as 6 (muntjac deer), but the number is fairly stable for each species. We know that chromosomes are made of protein and nucleic acid. The nucleic acid is deoxyribonucleic acid (DNA for short); the acid forms long chain molecules made up of phosphate and sugar sub-units. Attached to the sugar are nucleotide bases, which are of four kinds – adenine, cytosine, guanine and thymine. The complex double helix structure of the DNA molecule was first worked out in 1953 by Crick and Watson. The double helix model offers a simple method for replication or reproduction of the molecule, and it also offers a mechanism for carrying coded messages. And so with this knowledge it has been possible to crack the genetic code, made up of 64 possible combinations of the four bases. These combinations specify the proteins which influence the visible expression of the character – whether a horse will have black, white or brown hair, or combinations of these colours.

Each species has its own genetic code which it passes on to each new generation. The system is tightly locked

Archaeopteryx, the first bird. Upper Jurassic of Solnhofen, Germany. The fossil displays both reptilian characters (e.g. teeth, claws on the forearm, tail with vertebrae) and avian characters (e.g. feathers and also clavicles fused to form the furcula or wishbone).

and difficult to break. However, it can sometimes be broken, and when this happens we may see the appearance of a new species, if the new combination is viable. Few mammalian species can be traced back with confidence more than a few million years, and there are few mammalian genera more than 10 million years old. Over the eons of time species have been replaced by new species, better able to compete or more adapted to survive in other environments.

Darwin's work firmly established the concept that all forms of life are related and present-day forms are the product of descent with modifications from ancestral forms. Evolution seen in these broad terms has been fully accepted by the scientific community worldwide. Darwin's mechanism of natural selection has not been disproved, but it may not be the only mechanism and details of its working are still being analysed.

In Darwin's time two major difficulties seen by fellow scientists were the shortage of geological time and the problem of 'missing links'. Darwin thought of the earth as many hundreds, perhaps thousands, of millions of years in age. His time scale was not shared by the physicists who thought of the earth as a mere one or

two million years old. The problem could not be quantified until early this century with the discovery of the properties of radioactive minerals; these give dates which show that Darwin was thinking in the right order of time.

A logical consequence of Darwin's proposal that species change gradually, is that the fossil record should preserve the intermediate forms; where, for example, were the links between the fish and amphibians, the amphibians and reptiles, the reptiles and birds, and between the reptiles and mammals? In the 1850s our knowledge of the fossil record was slight, and that record is as we know imperfect. The vertebrate classes all stood quite distinct from each other; the links were all missing. Now over a hundred years later we have found many of these missing links. *Archaeopteryx* from the Jurassic of Germany is half reptile, half bird. We also know of a series of fossils that straddle the line between advanced reptile and early mammal (discussed in Chapter 4). So both of these difficulties with Darwin's theory have been overcome, and our discussions today centre on the detailed mechanisms of species origin.

Of the many models that have been proposed to

European hedgehog *Erinaceus*

European mole *Talpa*

The European mole (*Talpa*) and the European hedgehog (*Erinaceus*) are very different in appearance but their similar dental anatomy shows them to be closely related insectivores. However, the European mole and the Australian marsupial mole (*Notoryctes*), although they look very similar, are only distantly related.

Marsupial mole *Notoryctes*

account for the origin of new species, interest currently centres on two which we can name the gradualist and the punctuated equilibrium models. The gradualist model proposes that species originate slowly over a long period of time by gradual accumulation of small changes being handed down from one generation to the next. The punctuated equilibrium model proposes that species remain static over long periods of time, then relatively suddenly an isolated subset or peripheral population undergoes a salient genetic (and morphologic) change; if and when this neomorph is reunited with the main population it may in competition survive and even oust the older type, thus establishing a new species lineage. To test these models requires a virtually complete fossil record over several million years; such examples are rare. Most of the tests have used marine microfossils. The sea sediments have fewer breaks than continental sediments; microfossils are more plentiful than macrofossils and can be collected at closer vertical (time) intervals. Results to date are equivocal and it is not improbable that both mechanisms operate.

Relationships

A problem that has long confused naturalists is the extent to which similarity of form indicates degree of relationship. Medieval scholars, for example, grouped whales with fish – they look alike and both live in the sea; likewise bats have been grouped with birds. It was the British naturalist John Ray who in 1693 first clearly recognised the mammals as the natural group we know today. The actual name 'Mammalia' for the class dates from the 1758 work of Linnaeus.

Mammals we may define as vertebrates with hair in which the female possesses mammary glands with which to suckle the young. But even within the

mammals, there are many problems of relationships. For example, moles of Europe and Australia look alike; alike in size, head shape, stubby snout, vestigial eyes, strong front feet with large claws, short tail and very fine velvety fur. All these similarities might suggest a close relationship; one could even go further and note that the European mole was black and the Australian mole white; this might mean two geographic races. But they have other differences which taxonomists regard as fundamental. The Australian moles are pouched marsupials reproducing like all other marsupials (kangaroos, wallabies and opossums). The European moles have a placenta and reproduce like all other European mammals. This difference is seen to set the two moles very far apart, throwing their common ancestry back into the Mesozoic. At the other end of the spectrum we could take the European mole and European hedgehog. Moles and hedgehogs do not look much alike; both eat earthworms and both live in Europe. But more importantly both have similar patterns of reproduction, have similarities in the arrangement of their ear bones and in the cusps of their teeth; all this leads taxonomists to regard them as closely related. Scholars have for the past two centuries wrestled with problems of this kind, attempting to obtain a 'natural' classification; by this they mean a classification that accurately reflects phylogenetic relationships and evolutionary branchings.

A totally new approach to such problems is that taken by the German taxonomist Willi Hennig. His system is usually referred to as cladistic which simply means branching, though in this instance it refers to dichotomous branching, that is branching into two. Hennig proposed that we look at characters as either primitive (plesiomorphic) or derived (apomorphic). The technique used aims to identify groups of organisms

Monotreme mammals

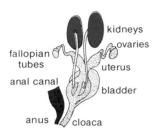

Openings from the large intestine (the anal canal), bladder and uterus all meet in the cloaca

Marsupial mammals

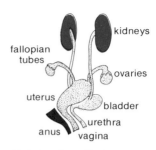

Openings from the bladder and uterus unite in the cloaca but the anal canal remains separate

Placental mammals

Openings from the large intestine (anus), bladder (urethra) and uterus (vagina) are all separate

Reproductive strategies define the three major groupings of mammals: the egg-laying monotremes, the pouched marsupials and the placental mammals.

sharing derived characters (synapomorphies); these will constitute monophyletic stocks, that is all the organisms that have arisen from a common ancestor. The cladist then searches for a 'sister' group, a group of organisms that branched off from the last common ancestor of a monophyletic stock. The cladistic approach has aroused much controversy, due in part to the unorthodoxy which leads to strange and unfamiliar groupings of animals. The methodology of cladistics is very rational and logical, but it does not often lead to a unique solution. Two factors play a large influence in the construction of cladograms, the dicotomously branching diagrams that illustrate relationships and out of which phylogenies may be derived. The first is the principle of parsimony; the fossil record being incomplete, it often happens that it is not possible to decide upon the time of origin of characters nor the direction in which they have changed. The polarity choice can be arbitrary and in cases of conflicting evidence the practice is to apply Occam's razor – that is, the simplest solution will be adopted. This parsimonious approach may not yield the right answer, but it is easiest to defend and likely to cause least errors. The second factor is the problem of convergence and parallelism. When are similar characters due to close relationship and when due to convergence from diverse ancestry? This problem arises in all taxonomic studies, cladistic or otherwise.

An example may serve to illustrate these problems. Let us consider the relationships of the cow, the lungfish and the salmon. Everybody would agree that as they all three have a vertebral column, a head with mouth and pair of eyes, muscles supplied with nerves and blood vessels, and so on, they all belong to the same phylum Chordata, or rather more loosely 'the vertebrates'. Within the vertebrates, all three animals could be seen either as equally related or as unequally related. If they are all equally related, we can say no more other than give them equal rank. If they are unequally related, then there are three possible ways in which two may be more closely related to each other than either is to the third species; the cladograms illustrated here show these alternatives. Each has its pros and cons and there is no intrinsic way of discovering the truth.

Living mammals fall readily into three major groups – the monotremes (egg-laying mammals such as the platypus), the marsupials (pouched mammals such as the kangaroo) and the placentals (such as dog and cat). When we take the fossil record into account the classification becomes more complex. The classical and the cladistic approach to mammalian classification produce broadly similar groupings of species, genera and families. But the grouping of the orders within the class Mammalia is keenly debated and no consensus has as yet emerged. The system presented here owes perhaps more to convenience and compromise than to confidence or correctness. This text does not rigorously follow the taxonomic groups, but is primarily based on trophic groupings, that is on the feeding specializations.

Cladist

On recency of common ancestry, first branch Early Devonian, second branch Middle Devonian; lungfish more closely related to cow than to salmon.

Anticladist

On shared skull roof characters, cow more closely related to salmon than to lungfish; no time axis implied.

Traditionalist

On degree of morphological divergence, the two fishes are more closely related than the mammal; no time axis implied in cladogram.

Cladograms to illustrate possible relationships of salmon, lungfish and cow.

Feeding types

There are three basic trophic groups in mammals as in reptiles; the insectivores, carnivores and herbivores, each with specialized offshoots and with gradations between. The term insectivore is used for small mammals that are usually opportunistic grubbers, feedings on worms, larvae and a wide variety of small invertebrates. Moles are classified within the order Insectivora, but feed largely on worms. The division between insectivore and carnivore is blurred, many small mammals having a mixed diet of insects and small vertebrates. On another side of the triangle many rodents have a mixed diet of grubs, larvae, seeds and fruit, thus straddling the line between insectivores and herbivores. Bears and pandas are mammals which

THE HIGHER CLASSIFICATION OF MAMMALS

Class	MAMMALIA	
Subclass	PROTHERIA	
Order		†Triconodonta
		†Docodonta
		Monotremata
		†Multituberculata
Subclass	THERIA	
Infraclass		†Pantotheria
		Marsupialia
		Eutheria
Cohort		Edentata
		Epitheria
Grand order		†Ictopsia
		Anagalida
		Ferae
		Insectivora
		Archonta
		Ungulata

†Extinct

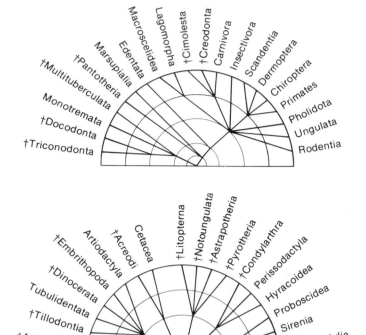

Top: a phylogeny of the mammals showing relationships of the major stocks. Above: details of the ungulate orders. †Extinct groups.

combine carnivorous and herbivorous items in their diets. Badgers, together with pigs and man, can be described as omnivores, so varied is their trophic intake.

Then there are the specialized offshoots with restricted diet, and usually this is associated with a high degree of adaptation to the diet; shell-crushing sea otters, fish-eating sealions, nectar-drinking marsupials, fruit-eating bats, anteaters and so on.

Most Mesozoic mammals were broadly insectivores and it is from this core that the other two trends of carnivory and herbivory evolved in several lineages and at different times and in different places. Thus as we shall see the taxonomic and trophic grouping do not always coincide. There are members of the order Carnivora that feed very largely on insects (aardwolf) or vegetation (giant panda). And there are carnivores among the marsupials, whales and the extinct creodonts.

Formal classification

The formal classification of mammals shown here considers a primary twofold division of the class Mammalia into the subclasses Prototheria and Theria. The prototherians are all extinct save for the egg-laying platypus and echidnas of Australasia. Prototherians

flourished in the Mesozoic, but few survived into the Caenozoic. The subclass Theria includes the vast majority of fossil and living mammals and has three major groups – Pantotheria, Marsupialia and Eutheria. The pantotheres are Mesozoic and include ancestors of all marsupials and eutherians. The marsupials or pouched mammals survive today mainly in South America and Australia. The eutherians or placentals comprise all other living mammals and many extinct stocks.

Within the Eutheria the first division is twofold into Edentata (armadillos, sloths and hairy anteaters) and Epitheria. The latter grouping contains some 32 orders of which about half are extinct. It is the grouping of these orders that is still much debated. The form tabulated here has a large measure of agreement if not universal assent. The first four grand-orders are all poorly known in the fossil record and only one has a living representative. The Ferae, Insectivora and Archonta each provide fairly readily recognized natural groupings. The Ungulata (hoofed mammals) is an exceedingly large grouping with about 21 orders, and requires further subgroupings to make it more manageable. The Rodentia stand apart from all other orders in an isolated grouping.

CHAPTER 3
Bones and teeth

Mammals are built of soft and hard tissues; with rare exceptions it is only the hard parts that are fossilized. Preservation of muscle, skin, blood vessels, nerves, brain or stomach require very special conditions and their chance occurrence in the fossil record is a notable event. Such are the frozen mammoths or the Messel mammals from West Germany.

The skeleton

The mammalian skeleton is composed of bone (calcium phosphate), is jointed and mostly internal. External parts include horn cores, antlers, nails, hooves, claws, and most importantly the dentition. Occasionally scales develop (pangolins and armadillos) and these may ossify as in the extinct glyptodonts. The main functions of the skeleton are to support the mammal, to serve as attachment areas for muscles which can contract and so move the body, and finally to protect vital areas such as the brain. The shape of the bones tell us a lot about how the animal moved, what sort of musculature it had; in some regions, especially in the skull, we learn about the blood vessels and the nerves passing through openings in the bones. Further the brain is close to the skull roof and leaves an imprint of its pattern which can be read.

The teeth are very characteristic of the family of mammals to which they belong, especially the cheek teeth; they enable us to identify accurately the species and to deduce what sort of food was eaten, how it was masticated and how old the animal was.

Examples of some of the varied uses of the mammalian tail. As a fifth limb for brachiating in a monkey, as a sunshade in a ground squirrel, as a paddle for swimming in a beaver and for fat storage in sheep.

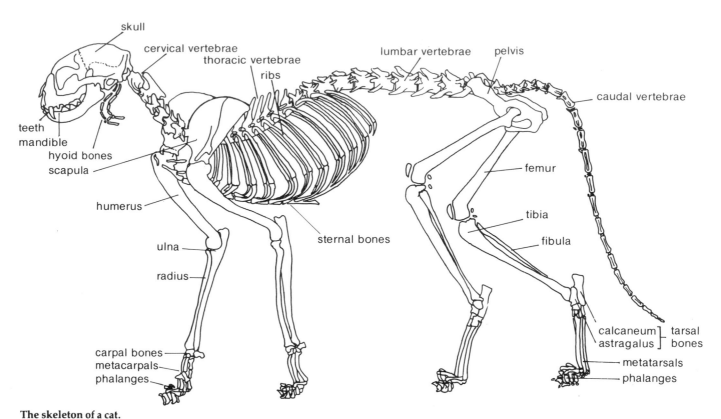

The skeleton of a cat.

Bones and joints

Mammal bones usually have a dense outer zone and a spongy inner zone; limb bones are hollow and in the core blood is produced. Bones articulate with each other by means of joints, which are encased by soft tissue and fluid filled. There are two basic types of limb joint, the pulley joint and the ball and socket joint. The pulley joint allows great freedom of movement in one plane and none in any other plane; examples are the finger and toe joints. The ball and socket joint is a universal joint allowing movement in any direction, such as seen in the shoulder and hip joints. Some joints, however, are combination joints; the elbow joint in man combines a pulley (humerus/ulna) with a ball and socket joint (humerus/radius). The knee joint is unusual in that in essence it comprises a sphere (femur) resting on a nearly flat surface (tibia); the passage of musculature across the joint is aided by an ossification, the knee cap or patella. The knee joint is a major load bearing joint in bipedal mammals and in man can give much trouble; it seems inadequately designed to say the least.

Vertebral column

The skull is joined to the vertebral column by a pair of hemispherical condyles, which allow great freedom of movement. The vertebrae usually have flattish articulations, each moving a small distance, but with many vertebrae moving together the back can be bent through a wide arc.

All mammals have a basically similar skeletal build – an axial and an appendicular skeleton. The axial skeleton comprises the vertebral column, the ribs, the sternal bones and the skull. The appendicular skeleton comprises the pectoral and pelvic girdles, the fore and hind limbs.

The vertebral column comprises a chain of vertebrae, each with a separate ossification articulating with the vertebra in front and behind. The column is divided into regions – the cervical or neck, the thoracic or chest, the lumbar or trunk, the sacral or hip, and the caudal or tail. In mammals the number of vertebrae in each region is fairly constant. Each vertebra comprises a bony disc or centrum which is the ossified notochord of primitive vertebrates. An arch rises above that, enclosing the nerve cord which runs from the brain all along the back. Above the arch rises the neural spine which carries the musculature that enables the back to be flexed and extended in a vertical plane. Issuing laterally from the base of the arch or the top of the centrum are a pair of transverse processes; attached to these are muscles which allow the animal to bend in a horizontal plane. The vertebrae articulate with each other by means of articular facets which arise from the neural arch. There are usually seven cervical vertebrae; the first and second vertebrae are always specialized with ball and socket joints to allow the head freedom of rotation. The thoracic vertebrae are characterized by the attachment of ribs, one pair to each vertebra; there are usually 12–15 pairs of ribs. The anterior thoracic vertebrae often have very long neural spines for the attachment of muscles from the back of the skull. The articular facets on thoracic vertebrae are so placed that there is usually great freedom of movement, not only in a vertical and horizontal plane, but also spirally, enabling mammals to twist their backs. This twisting action is unknown in reptiles; it gives the mammal greater flexibility for climbing, for turning, for grooming and most importantly it enables a female to sit down, lie on her side and suckle her young. Reptiles cannot lie on their sides, and to take the weight off their legs they just let their bellies rest on the ground. The lumbar vertebrae usually number five or six; they characteristically have long transverse processes to which are attached the heavy muscles from the hip region. There is little freedom of movement in the lumbar region and emphasis is on strength and support rather than on flexibility. The succeeding sacral region normally comprises three fused vertebrae to which are attached the pelvic bones. The caudal or tail region is the most variable, ranging from a few poorly formed vertebrae as in man to the long prehensile tails of monkeys, or the strong support tails of kangaroos. Tails in mammals are used for a variety of purposes: for swimming in whales, otters and beavers; for hanging from branches in monkeys, pangolins and some mongooses; for balance in many arboreal species and in fast running species like leopards and cheetahs; for fat storage in sheep; as flywhisks in horses; as a sun-shade in ground squirrels; as a heat insulator in cats and other small carnivores; for the attachment of flight membrane in bats, and many other uses.

Ribs do not often preserve well; they are readily broken and are difficult to identify. Rib cages are long in many small carnivores, seals and sealions which have large lungs. They are short and broad in many primates, where the movements of the diaphragm rather than the ribs are more important in expanding the lungs. There are some specializations such as the imbricating or overlapping ribs of some arboreal edentates and primates for increased stability. Mammals which rest on their bellies, such as sirenians, seals, sealions and walruses, tend to have very strong sternebrae, which are a short series of breastplate bones and serve to anchor the cartilaginous extensions of the ribs on the ventral surface.

Limbs and girdles

Turning now to the appendicular skeleton, both the fore and hind limb are constructed on similar patterns, that is a girdle bone, single bone first segment, double bone second segment, and a third segment comprising a series of small bones, primitively with five digits.

The scapula or shoulder blade is not fused to the vertebral column but held in place by muscles only; this allows considerable freedom of movement as seen when a lion stalks. There may be a second pectoral bone, the clavicle or collar bone; this is present in arboreal mammals and helps to prevent dislocation of the shoulder in climbing activities. The clavicle is lost in many mammals, especially running types like horses and antelopes. The innominate which forms the pelvic girdle is firmly joined to the sacral vertebrae; this union, however, rarely amounts to total fusion since some expansion of the pelvic girdle is required during parturition; most of this movement is at the junction of the two pubic bones, but if the sacral joint were wholly

ossified this would be impossible.

Limb lengths vary a great deal. Bipedal rodents (kangaroo rats) have long hind legs, brachiating gibbons have long front legs, whales have lost their hind legs and bats have very long fingers on the front limbs. The variations all reflect the multifarious uses to which the limbs are put – walking, running, climbing, flying, swimming, hopping, as well as digging, catching prey, holding food, carrying young, grooming and fighting. When power is required (for digging or swimming), the proximal bony elements (humerus and femur) are usually short and the muscles so arranged that their mechanical leverage has a long moment arm ensuring powerful flexion. When speed is required, as in horses, these same bones are elongated and hence muscles with short moment arms ensure that the flexion will be rapid and extensive. In almost all hoofed mammals (ungulates) the paired leg bones of the second segment are fused, so that there is no possibility of rotating the hand or foot; this is an advantage in that it lightens the weight to be moved and allows the excess musculature and ligaments that would be required to keep the foot in a forward position to become vestigial. Also in ungulates the number of toes is usually reduced, to two in cattle and antelopes, and to one in horses. The hoofs develop as shock absorbers to take the impact of the body weight when galloping. Carnivores usually retain the full five digits and develop prominent claws for grasping their prey.

Skull

The final area of the skeleton to be examined is the skull and mandible. The head is the centre of sensory perception, of smell, sight and hearing. It also houses the centre of the neural control – the brain. Food is taken in through the mouth and masticated by the teeth, which are set in the jaws and operate by the contraction

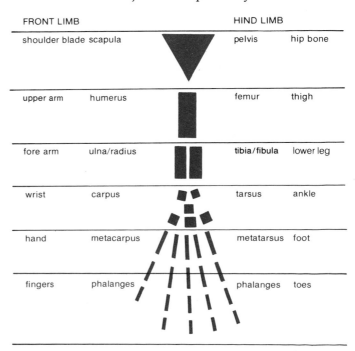

FRONT LIMB			HIND LIMB	
shoulder blade	scapula		pelvis	hip bone
upper arm	humerus		femur	thigh
fore arm	ulna/radius		tibia/fibula	lower leg
wrist	carpus		tarsus	ankle
hand	metacarpus		metatarsus	foot
fingers	phalanges		phalanges	toes

The bones of the limbs in a typical mammal.

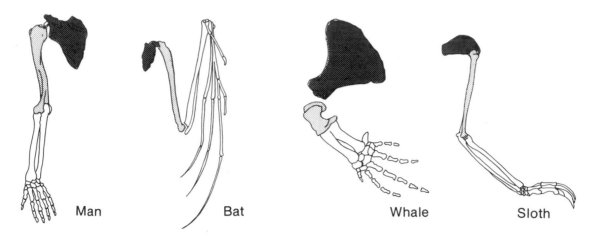

Some examples of adaptation of the mammalian fore limb. Grasping (man). Flying (bat). Swimming (whale). Brachiating (sloth).

Comparable elements can be traced through the series; the changes being normally in proportion and shape.

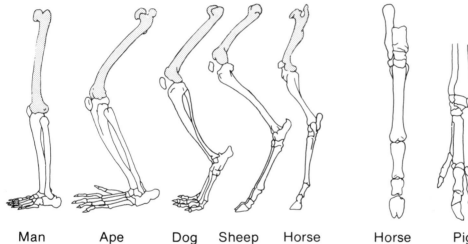

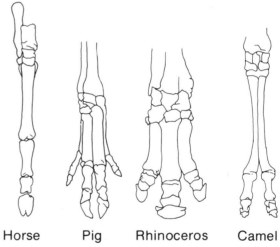

Some examples of modifications in the mammalian hind limb. Plantigrade (man). Plantigrade (ape). Digitigrade (dog). Unguligrade, two toed (sheep). Unguligrade, one toed (horse). The series shows progressive elongation of the foot bones from left to right, which correlates with faster running speeds.

Examples of foot adaptations and toe reduction from the primitive five toed condition. One toed horse. Four toed pig with side digits reduced. Three toed rhinoceros. Two toed camel with splayed digits.

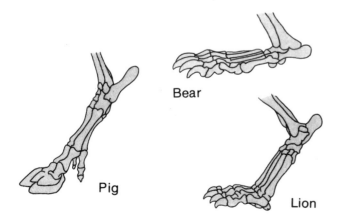

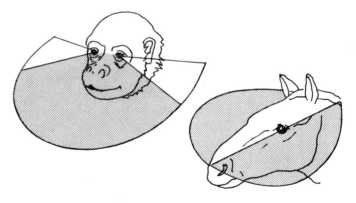

Three foot postures of mammals.
Plantigrade as in bear. Digitigrade as in lion. Unguligrade as in pig.

Stereoscopic vision is achieved in most arboreal mammals and many carnivores. In these mammals the eyes are placed side by side on the front of the face. In most herbivores the eyes are placed laterally which gives good all round vision. The examples are of a monkey and a horse.

of powerful jaw muscles. The head region is further the centre for attack and defence, using large canine teeth, or bony outgrowths such as antlers and horns. The ultimate shape of the skull and mandible is the byproduct of all these competing requirements, and since these needs are different in each species, the end product is distinct for each species.

The brain is virtually never preserved intact, but in two ways we can learn almost all the details of its surface anatomy. The skull on burial may fill with fine sediment, which on compaction gives a natural brain cast. Alternatively if the skull is free of matrix, then a rubberoid plastic can be used to produce an artificial brain cast. Both give detailed information on the contours of the brain, its volume and the position of the cranial nerves.

SENSES The nasal region houses the olfactory or smell sense; in some mammals such as anteaters, deer and dogs it is highly developed, while in others such as whales, otters and sealions it is poorly developed. Study of the proportions of the nasal region, the internal nasal bones and the olfactory lobe of the brain enable us to make deductions about its acuity. Vision is common to almost all mammals save a few burrowing forms, though in some it is poor (elephants and whales). The most acute vision is found in mammals that climb (primates) and run at high speed (cheetahs and lions). They have stereoscopic vision to enable them to judge distance accurately and to home in on fast moving prey. The ear is the centre of balance as well as the centre of hearing. While the outer pinna or ear flap is not fossilized it is usually possible to determine from the structure of the middle ear whether the auditory sense was well developed. Desert living mammals tend to have the best hearing; this is partly because sound travels best in warm dry air, and because these conditions are poor for olfactory sensing. Further, desert mammals prefer to feed and hunt in the cool of the evening or early morning when the light is poor and sight is of little value. The ear region has another characteristic of great value; it has a very complex bony structure and through a series of foramina or holes pass the arterial blood vessels to the brain. The passages of these vessels is standard for each family of mammals and so the region is of great value in classifying mammals.

JAWS The two most important muscles used in closing the jaws are the temporal and the masseter. The temporal produces a powerful upward and a backward action and is best developed in carnivores. The masseter produces an upward and forward action and is best developed in herbivores.

TEETH In mammals there are two sets of teeth during a life-time – a milk or deciduous set and a permanent or adult set. If we take as an example the dentition of the dog we see a type of dentition not very different from that of primitive mammals. The total complement of permanent teeth is ten in each half of the upper dentition and 11 in each half of the lower dentition, making 42 in all. The dentition is divided into regions, which from front to back are incisor, canine, premolar

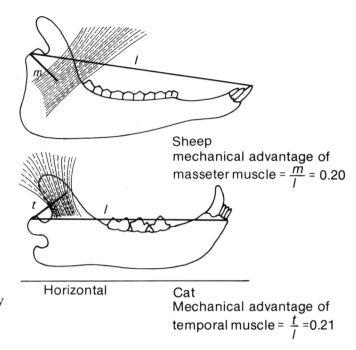

Sheep
mechanical advantage of masseter muscle = $\frac{m}{l}$ = 0.20

Horizontal

Cat
Mechanical advantage of temporal muscle = $\frac{t}{l}$ = 0.21

Examples of the jaw mechanics of a herbivore (sheep) and carnivore (cat). In the sheep the articulation is elevated above the dentition to obtain maximum mechanical advantage for the masseter, the main jaw-closing muscle in herbivores. In the cat the articulation is level with the dentition to obtain the maximum mechanical advantage for the temporal, the main jaw-closing muscle in carnivores.

and molar. There are three incisors in each half jaw; they are small chisel edged and single rooted teeth; their milk precursors are similar but smaller. Their function is to hold and tear. Behind these come the canines, one in each jaw half. These are large pointed tusks with single roots and are used for piercing; the milk canines are small scale versions. Then come four premolar teeth in each jaw half; the anterior one is small and if present at all is, in the mandible, a persistent milk with single root and single cusp. The second is usually double rooted with a more elongated crown, and the third premolar is an enlarged version of the second. The fourth premolar in the lower jaw is a yet bigger version of P_3 but in the upper jaw it is highly specialized. The tooth has three roots and the crown has an inner anterior cusp and an elongate blade on the outer border. This blade cuts against the first molar in the lower jaw, which comprises simply an elongate blade. These two specialized teeth, P^4 and M_1 are known as the carnassial or meat-eating teeth and with them the dog cuts meat into swallowable chunks. Behind these the other molar teeth have expanded surfaces which are used in chewing. In the milk dentition there are no molar teeth and the carnassial action is performed wholly by dP^3 and dP_4 (see p. 22).

Diets and dentition

We can thus classify mammals into trophic groups which will reflect their dental patterns. The carnivores such as the dog have a carnassial dentition and a preference for meat in their diet. The herbivores feed largely or wholly on vegetation and have usually lost or

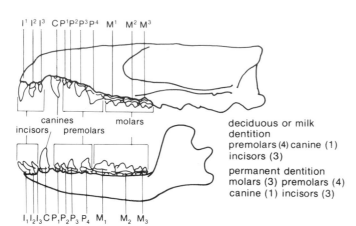

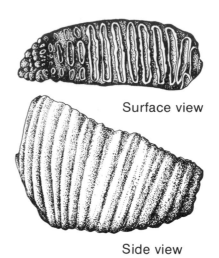

Surface view

Side view

The naming and numbering of permanent teeth in a placental mammal. Deciduous or milk teeth are the precursors of the incisors, canines and premolars. Molar teeth have no milk precursors.

The elephant tooth is an example of adaptation to grinding rough food, with numerous ridges of hard enamel across the grinding surface and a high crown; these characters reduce the rate of wear and lengthen the life of the tooth.

greatly reduced their canines; their cheek teeth are broad and relatively flat for crushing and grinding vegetable matter. Rodents have basically similar cheek teeth, but their incisors are specialized as efficient chisels for cutting tough fibrous material; they also have like herbivores a gap between the incisors and the cheek teeth, because the two parts of the dentition operate at different times and with different actions. Then there are the omnivores, which eat a wide variety of foods; like ourselves they have a complete dental battery without the specializations found either in carnivores or herbivores. Some other special types found are molluscivores with very thick crowns on the teeth (e.g. sea otter and walrus), frugivorous mammals with sharp pointed teeth (as fruit bats), insectivores and ant-eaters with either sharp pointed teeth or none at all, and piscivores (like seals and sealions) with simple pointed teeth to hold fish.

Mammal teeth have to withstand a life-time of heavy wear and are in consequence very strong. Three materials are involved in their structure. Dentine forms the bulk of the crown of the tooth; it is tough and can

withstand stresses in all planes without splitting; elephant tusks are wholly dentine – this is the major source of commercial ivory. The crowns of the teeth are capped with a skin of enamel, an even harder material and very disease resistant. Enamel is composed of minute apatite crystals (calcium phosphate); these make up 98 per cent of the enamel, the remaining two per cent being made up of a protein bonding. Dentine is about 70 per cent apatite with a collagen and water. By combining inorganic salts with organic compounds, bones and teeth achieve great hardness with low brittleness; both are as strong as mild steel, though not quite so stiff.

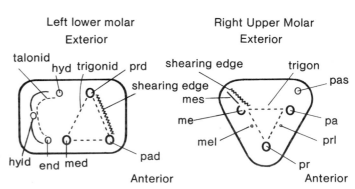

end   entoconid
hyd   hypoconid
hyld hypoconulid
me   metacone
med   metaonid
mel   metaconule
mes   metastyle

pa   paracone
pad   paraconid
pas   parastyle
pr   protocone
prd   protoconid
prl   protoconule

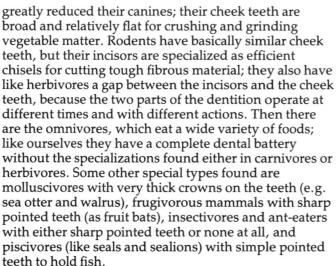

cement   enamel   dentine

The occlusal view of a horse tooth to show the pattern of development of enamel, dentine and cementum on the chewing surface.

The names of the main cusps on the upper teeth end in -cone, the minor cusps end in -conule or -style. Similar terms are used for the lower teeth, but all have a final ending -id.

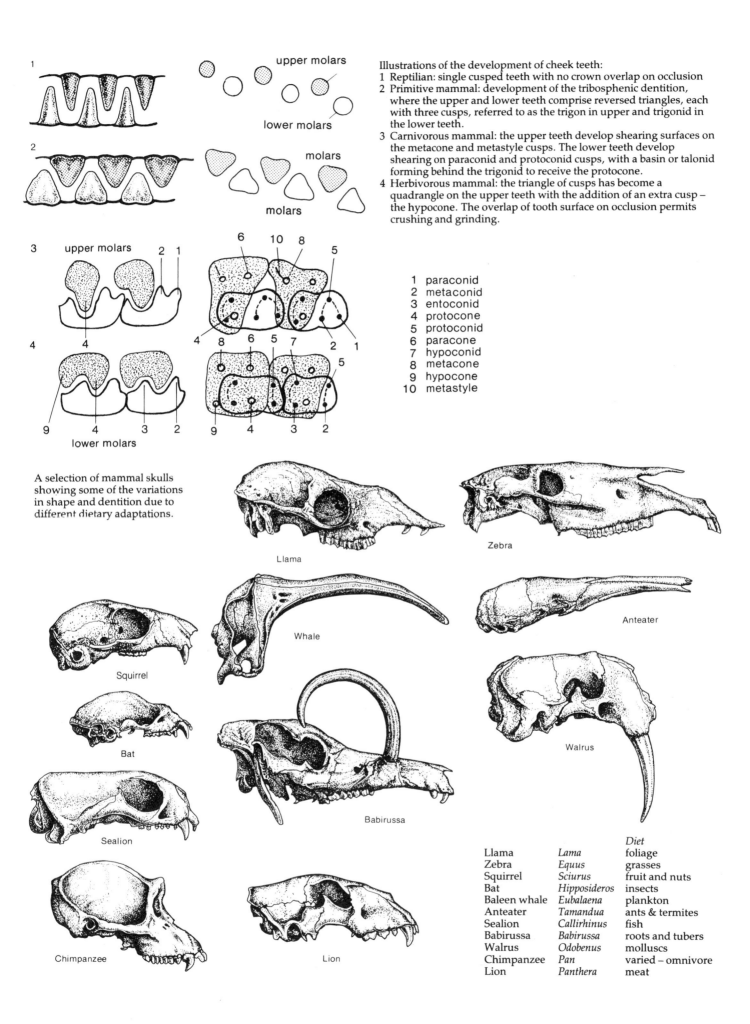

1 upper molars
lower molars

2 upper molars
molars
molars

3 upper molars
2 1
6 10 8 5
4 4 8 6 5 7 2 1 5
9 4 3 2 9 4 3 2
lower molars

Illustrations of the development of cheek teeth:
1 Reptilian: single cusped teeth with no crown overlap on occlusion
2 Primitive mammal: development of the tribosphenic dentition, where the upper and lower teeth comprise reversed triangles, each with three cusps, referred to as the trigon in upper and trigonid in the lower teeth.
3 Carnivorous mammal: the upper teeth develop shearing surfaces on the metacone and metastyle cusps. The lower teeth develop shearing on paraconid and protoconid cusps, with a basin or talonid forming behind the trigonid to receive the protocone.
4 Herbivorous mammal: the triangle of cusps has become a quadrangle on the upper teeth with the addition of an extra cusp – the hypocone. The overlap of tooth surface on occlusion permits crushing and grinding.

1 paraconid
2 metaconid
3 entoconid
4 protocone
5 protoconid
6 paracone
7 hypoconid
8 metacone
9 hypocone
10 metastyle

A selection of mammal skulls showing some of the variations in shape and dentition due to different dietary adaptations.

Llama
Zebra
Whale
Anteater
Squirrel
Bat
Sealion
Babirussa
Walrus
Chimpanzee
Lion

		Diet
Llama	*Lama*	foliage
Zebra	*Equus*	grasses
Squirrel	*Sciurus*	fruit and nuts
Bat	*Hipposideros*	insects
Baleen whale	*Eubalaena*	plankton
Anteater	*Tamandua*	ants & termites
Sealion	*Callirhinus*	fish
Babirussa	*Babirussa*	roots and tubers
Walrus	*Odobenus*	molluscs
Chimpanzee	*Pan*	varied – omnivore
Lion	*Panthera*	meat

Reconstruction of *Smilodon* head. Note the wide gape with the mandible swung far back to allow the sabre canines to be used for stabbing.

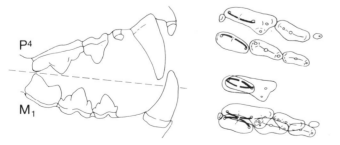

Diagram to illustrate the carnassial or meat slicing action of the posterior pair of cheek teeth in a cat. A blade is developed on P^4 and M_1 and the jaw closure is comparable to the action of a pair of scissors.

In mammals that eat molluscs and bones the enamel is very thick; examples are the sea otter and hyaena. The roots of teeth are formed of cement, which is intermediate in hardness between dentine and enamel. In some herbivorous mammals with diets that wear down teeth rapidly, as is the case with many rodents, true roots never form; the teeth remain open rooted and continue to grow throughout life, replacement growth balancing the rate of wear. In others such as elephants, the spaces between the cusps and lophs on the crown are infilled with cement which greatly decreases the rate of wear on the tooth as a whole.

It is perhaps ironic that humans struggle throughout their lives to hold on to their teeth, fighting off the ravages of caries and other hazards, yet after death the teeth outlast all other parts of the body. The soft tissues decay rapidly, the bones crumble without burial in a suitable medium, but the teeth survive most natural hazards to emerge as the commonest evidence of mammals in the fossil record.

CHAPTER 4
Reptiles into mammals

Lizard: a typical reptile. Note the absence of external ears, the scaly body and the sideways projecting limbs.

A lizard and a cat are clearly distinct and different animals, the first is a reptile and the second a mammal. We could also note that they shared many features in common – both have four legs, a tail, head, jaws, eyes and tongue. If we were to dissect them we would find that both had an articulated bony skeleton, a heart, blood, kidneys, brain, lungs and so on. They are related to each other in having a common ancestry back in the early Mesozoic. It was from reptiles that the mammals evolved during mid or late Triassic times.

To decide when a reptile became a mammal is difficult, as advanced reptiles gradually acquired more and more mammalian features. For convenience we can group the differences between the two orders as skeletal and non-skeletal. Reptiles have a low metabolic rate, generally move rather awkwardly with limbs that project sideways from the body, have variable body temperature and warm up only by exposing their scaly bodies to the sun; they lay eggs and have small brains

Domestic cat: a typical mammal. Note the prominent ears and the long furry tail which can be used wrapped round the body to help retain heat when asleep.

relative to their size. Mammals have a high metabolic rate, are able to maintain their bodies at a constant temperature and use their fur to help keep them warm; they move efficiently with limbs tucked directly under the body. Embryos of true mammals develop within the mother and are born at an advanced stage and are then suckled on milk; they have large brains relative to their size.

While all these features appear to differentiate a

reptile from a mammal, none of them fossilizes and in almost all cases there are exceptions to the rule. Pangolins are scaly mammals, bats allow their temperature to drop below freezing point, echidnas lay eggs. There is evidence that some pterosaurs (flying reptiles) had fur, that dinosaurs were warm blooded and that in ichthyosaurs (whale-like reptiles) the embryos developed internally and were born alive. There appears to be only one character that is totally unique to all mammals – the possession of specialized milk producing glands, the mammary glands, the source of the name mammal. No reptile, bird or other vertebrate possesses

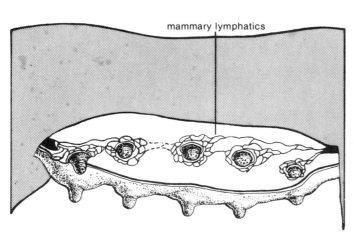

A reptile adopts a sprawling gait with the limbs emerging horizontally from the body. A mammal adopts an upright stance with the limbs placed directly beneath the body; this is mechanically much more efficient.

The arrangement of the mammary glands in a typical mammal showing the internal lymphatic connections.

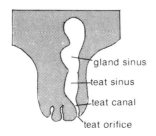

Section through a mammary gland. These milk-producing glands are modified sweat glands and characterize the female of every mammal species.

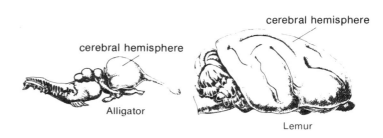

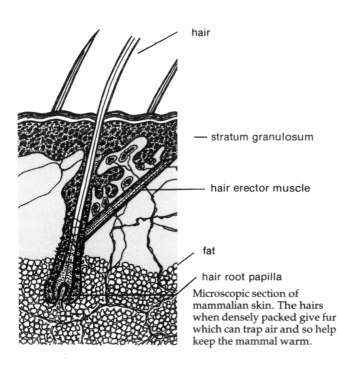

Microscopic section of mammalian skin. The hairs when densely packed give fur which can trap air and so help keep the mammal warm.

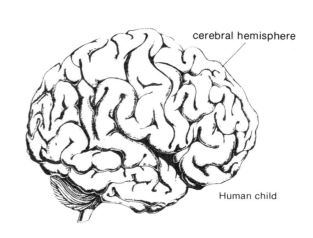

A reptile brain is small and simple compared with that of a mammal. In particular the anterior brain (cerebral hemispheres) expand in mammals, become convoluted and in man extend backward to overlap much of the remainder of the brain.

the ability to feed its young on mother-produced milk. While direct evidence of lactation does not fossilize, we can indirectly make deductions. Milk is very nourishing, very consistent in composition and always available 'at home'. The young have only to suckle and sleep; in consequence they grow very rapidly. Teeth are not

maturity approached the epiphyses become fused to the shaft. The thoracic vertebrae have articulations that enable a mammal to twist its back and so lie on its side, and suckle. The skull has a double condyle articulation with the vertebral column for greater freedom of movement, the hard palate is elongated to separate the

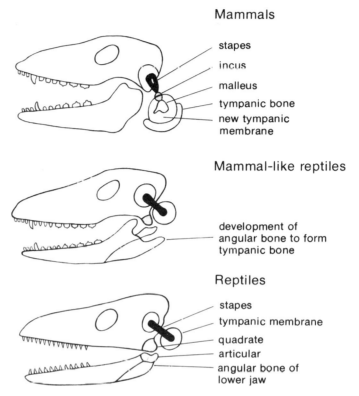

Section through the snout of a mammal. Air entering the nostrils first circulates around the anterior turbinal bones where it is filtered and warmed, then passes over the posterior turbinal bones which are innervated from the olfactory lobes of the brain and can thus sense smell. The secondary hard palate (absent in reptiles) separates the air passage from the mouth, thus allowing a mammal to breathe and eat simultaneously.

These diagrammatic skulls show the changes in the jaw articulation and ear region in the transition from reptile to mammal. Reptiles have single cusped teeth, the lower jaw has several bones and there is only one ear ossicle (stapes). In mammals the teeth are regionally differentiated, the lower jaw has only one bone and the other bones have taken on new roles in the ear region.

required until the young are weaned – indeed they would be a hazard if they erupted earlier; the skull and jaws can thus develop more fully before they appear. The teeth have numerous cusps and so in closing the jaws the upper and lower teeth occlude very precisely to achieve maximum efficiency. Teeth fossilize extremely well being composed of tough enamel and dentine, and their cusp pattern is usually highly diagnostic of the family, genus and often the species. The milk or deciduous set lasts until the mammal skull is nearly fully grown and is then replaced with the permanent set which has to last for the rest of its life. Some of the most advanced reptiles have limited replacement and multi-cusped teeth like those of mammals.

The skeleton also reflects physiological differences between the two classes; for example, the bones have limited growth and several centres of ossification. During growth the ends of the bones (epiphyses) are separated from the shaft by a layer of cartilage; as

wind pipe from the food in the mouth and so enable a mammal to breathe normally while chewing. But again any one of these characters can be found in transitional forms and no one appears to be uniquely characteristic (autapomorphic). Consequently we are left with only one group of osteological features associated with the jaw joint and middle ear on which we can rely for separation of the two classes.

In all reptiles the skull and mandible articulate using the quadrate and articular bones, and there is a single middle ear bone – the stapes. In all mammals the skull and mandible articulate using the squamosal and dentary bones, and there are three middle ear bones – the stapes, incus and malleus. The incus and the malleus are not newly acquired characters (neomorphs); they are the transformed quadrate and articular of the reptile which have changed their position and function. While there are fossil reptiles which approach the mammalian condition in jaw articulation characters, the one or three ossicles stands as a demarcation that can be applied to the fossils, if sufficiently well preserved for these delicate structures to survive.

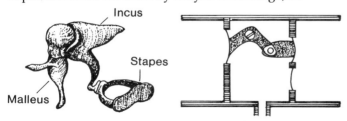

All mammals have three ear ossicles in the middle ear; these tiny bones act as levers to transmit sound waves from the ear drum to the inner ear. The ossicles shown are human and the ear mechanism is also shown diagrammatically.

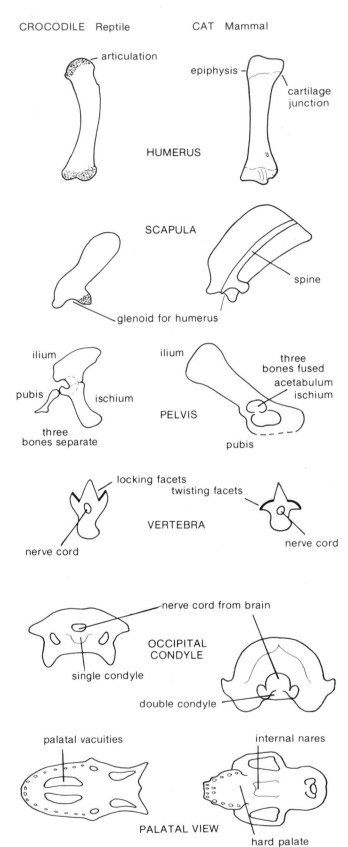

CROCODILE Reptile   CAT Mammal

HUMERUS

SCAPULA

PELVIS

VERTEBRA

OCCIPITAL
CONDYLE

PALATAL VIEW

Skeletal characteristics which differentiate reptiles and mammals.

PRINCIPAL DIFFERENCES BETWEEN REPTILES AND MAMMALS

REPTILE	MAMMAL
Non skeletal characters	
Epidermis scale covered	Epidermis hair covered
No external ear	External ear or pinna present
Oviparous (egg laying)	Viviparous (young develop within the mother)
Growth continues throughout life	Limited growth period
Sprawling gait	Erect gait
Ectothermic (cold blooded)	Endothermic (warm blooded)
Three chambered heart	Four chambered heart
Low metabolic rate	High metabolic rate
No diaphragm	Diaphragm present
Small brain	Large brain
Skeletal characters	
Bones without epiphyses	Bones with epiphyses
Ribs present on all vertebrae	Free ribs confined to the thoracic vertebrae
Scapula without spine	Scapula with spine
Pelvic bones separate	Pelvic bones fused
Skull with one occipital condyle	Skull with two occipital condyles
Bony palate incomplete	Secondary bony palate present
Paired external nares	Single external nares
Mandible composite	Mandible single bone (dentary)
Middle ear with one ossicle (stapes)	Middle ear with three ossicles (stapes, incus and malleus)
Jaw joint formed on quadrate and articular bones	Jaw joint formed on squamosal and dentary bones
Homodont dentition	Heterodont dentition
Teeth continually replaced	Two dental sets only, deciduous and permanent

The early reptiles

Towards the end of the Carboniferous period, some 300 million years ago, the first true reptile stock appeared. Amongst the variety of reptilian forms which evolved at that time, two forms were to become particularly important – the synapsids and diapsids. The synapsids were a group that developed mammal-like tendencies whilst the diapsids were to remain largely reptilian, but were to give rise eventually to lizards, crocodiles, dinosaurs, birds and flying reptiles. The synapsids became the dominant reptile stock of the Permian and early Triassic, but declined virtually to extinction with the rise in the early Jurassic of the diapsid group in the form of dinosaurs. Rather strikingly, we must conclude that the evolution of mammals suffered a dramatic change of fortune, almost becoming extinct soon after they arose. However, these synapsid descendants were eventually to triumph on land after the extinction of the dinosaurs, flourishing throughout Caenozoic times as the dominant vertebrates. The diapsids have left their descendants too; direct inheritors of the dinosaurs are the birds, whilst other diapsid reptilian forms continue to flourish as lizards, snakes and crocodiles.

The mammal-like reptiles varied enormously in size from small lizard-like forms to creatures the size of a rhinoceros; their dietary specializations included insectivores, herbivores and carnivores, evolving many features that were later to be paralleled by the mammals. Nonetheless, they remained essentially a conservative group; there are no bipedals, flying or truly aquatic forms. Their most characteristic feature, and that to which they owe their scientific name of Synapsida, is the presence in the skull roof of a single opening for the attachment of jaw muscles. The diapsids have two such openings. A second distinctive feature was their experimentation with temperature control. Living reptiles are poikilothermic or cold blooded, body temperature varying with that of the environment. In order to be active, they sun themselves and warm up their bodies; in other words, they can raise their body temperatures only from external sources – an ectothermic control. Similarly, they can only lower their body temperature by seeking shade. This type of control became more precise in the case of some species; the sail of *Dimetrodon* was probably highly vascularized (like elephants' ears) and would have enabled the animal to warm up rapidly and so have a head start in searching for its prey, a more sluggish herbivorous species. In reverse, it could use the same system to cool down, and by cutting off the blood supply to the sail it could retain a high temperature when the ambient air temperature fell in the evening. This example of primitive temperature control was to foreshadow more complex forms in later Triassic species.

The synapsids, reptiles with mammalian aspirations

Pelycosaurs

The synapsids comprise two orders, the primitive pelycosaurs and the advanced therapsids.

Pelycosaurs are well known from the early Permian deposits of Texas. They first appear in the late Carboniferous and had become extinct before the close of Permian times. Few are known outside North America, but some remains have been found in Europe and perhaps South Africa. *Archaeothyris* was a small, latest Carboniferous, insectivorous, lizard-like form. *Varanosaurus*, from the early Permian was about a metre in length, including its very long tail; it had an elongate snout with canine-like tusks and appears to have been semi-aquatic, probably eating fish. *Dimetrodon* is an example of a varied and highly successful stock of carnivorous sailed-lizards. The jaws had a battery of

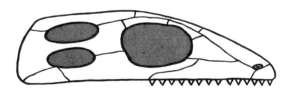

Two types of reptile skull. Top: the synapsid skull with single opening behind the orbit for the insertion of jaw muscles. Bottom: the diapsid skull with two such openings.

Varanosaurus was about 1 m long; the hind legs were longer than the front legs, and this combined with the elongate snout and battery of small teeth suggest a fish-feeding semi-aquatic adaptation.

Dimetrodon was about 3 m long. The narrow body, elongated limbs and relatively light tail all suggest it could have moved rapidly. The head is large with a formidable dentition to feed on other reptiles and amphibians. The sail was high and probably acted as a recognition signal as well as a thermal regulator. It is likely that the skin was at least partly keratinized so that it could stay out of water for considerable periods. Early Permian of Texas.

sharp pointed teeth, some of the more anterior ones being canine-like. The larger species reached over 3 metres in length and may have preyed on smaller species as well as on their herbivorous relatives. *Edaphosaurus* with its sail looks superficially like *Dimetrodon*, but the skull is smaller, has no canine teeth and the cheek teeth are blunt suggesting a herbivorous diet. We can follow the change from early semi-aquatic pelycosaurs living on equatorial deltas to later, larger, and more terrestrial forms living in a drier climate with only seasonal rainfall. It was from these later pelycosaurs that the therapsids were to arise.

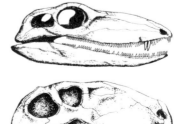

Left, top: skull of primitive pelycosaur *Varanosaurus* (15 cm long); the elongate snout and jaw are armed with a battery of small sharp pointed teeth and a pair of incipient canines. *Varanosaurus* probably fed on fish. Below: skull of the carnivorous pelycosaur *Dimetrodon* (30 cm long); the teeth are heavy and pointed with well-developed canine tusks. (After Romer & Price).

Restoration of *Edaphosaurus* among the vegetation of the Texas deltas in Early Permian times (260 MYA). This herbivorous reptile could use its sail to keep cool, as elephants use their ears today.

The 3 m long *Edaphosaurus* had a heavy tail, short stout legs, very short neck and small head. The vertebral spines are knobbly and in life were covered with well vascularized skin and used to regulate temperature. (After Romer & Price).

Skull of *Edaphosaurus* (15 cm long) in lateral and palatal view. Note the large eye socket and the jaw articulation which is placed low down near the back of the mandible. The teeth are small and blunt, without canine tusks and palatal teeth help shred vegetation. (After Romer & Price).

Skull of semi-aquatic dicynodont *Lystrosaurus* from the early Trias. The downturned snout and well-developed tusks may be adaptations to bottom feeding. (After Broom).

Lystrosaurus was one of the most successful of synapsid reptiles; it has been found in South africa, India, Antarctica, China and USSR, where it lived in rivers and lakes during the early Triassic.

Above, top: skull of *Titanophoneus* (40 cm long), a carnivorous dinocephalian from the mid Permian. Bottom: skull of *Ulemosaurus* (40 cm long), a herbivorous dinocephalian from the late Permian. Both genera are from the Ural mountains of USSR. (After Orlov & Efremov).

Skull of *Dicynodon* (20 cm long) from the later Permian of South Africa. This herbivore had a beak-like jaw with horny pads and a well-developed pair of tusks. (After Broom).

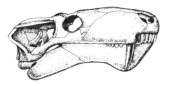

Above, top: skull of *Lycosuchus* (25 cm long), a therocephalian carnivore from the mid Permian of South Africa with two large canine tusks. Bottom: skull of *Scymnognathus* (30 cm long), a gorgonopsid from the late Permian of South Africa. Gorgonopsids were the dominant carnivores of the period and their sabre-like canines were protected from damage by a deep mandible. (After Broom and Watson).

Scene in the late Triassic 220 million years ago.

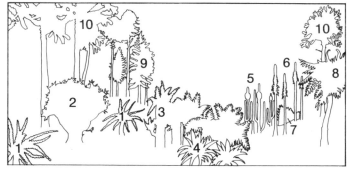

■ Key to illustration
1 *Neuropteridium*   2 *Caytonia*(Seed fern)
3 Cycadeoids   4 *Pelourdea(Yuccites)*
5 *Equisetites*   6 *Pleuromeia*
7 *Equisetum*(Horsetail)   8 *Itopsidema* (Tree fern)
9 *Brachyphyllum*   10 *Araucarites*(Monkey puzzle)

Therapsids, direct ancestors of the mammals

The therapsids had a much wider distribution ranging over the globe, including Antarctica, and persisting for some 80 million years. About 36 families and very many genera of therapsids are known. The classification of them is complex and we will here discuss only the three major suborders – dinocephalians, dicynodonts and cynodonts – with a passing mention of three other minor suborders – gorgonopsids, therocephalians and tritylodonts.

Dicynodonts were very abundant in the late Permian – 90 per cent of all reptile skulls found in South Africa at this time are dicynodont. The skull is highly modified to a herbivorous diet and its most striking feature is the almost total absence of teeth, the upper canine tusks being the only prominent survivors, and giving the group its name. Tusks are not present on all individuals and may have been better developed in males. In place of cheek teeth the jaws carried horny pads as do turtles; some pads were flat for crushing and others sharp edged for cutting vegetation. *Dicynodon* is known from about 100 species in the later Permian. The most interesting Triassic form was *Lystrosaurus*; it had poorly ossified feet, short limbs and short tail, and appears to have been semi-aquatic. It had a very widespread distribution, being known from South Africa, Antarctica, India, USSR and China. *Kannemeyeria* was a large heavy cow-like dicynodont known from the early Triassic of South Africa, South America and Russia.

Scymnognathus was an abundant gorgonopsid in the late Permian of South Africa. It was a large, lightly built and long limbed carnivore with a short snout and very big canine teeth resembling the sabreteeth which later

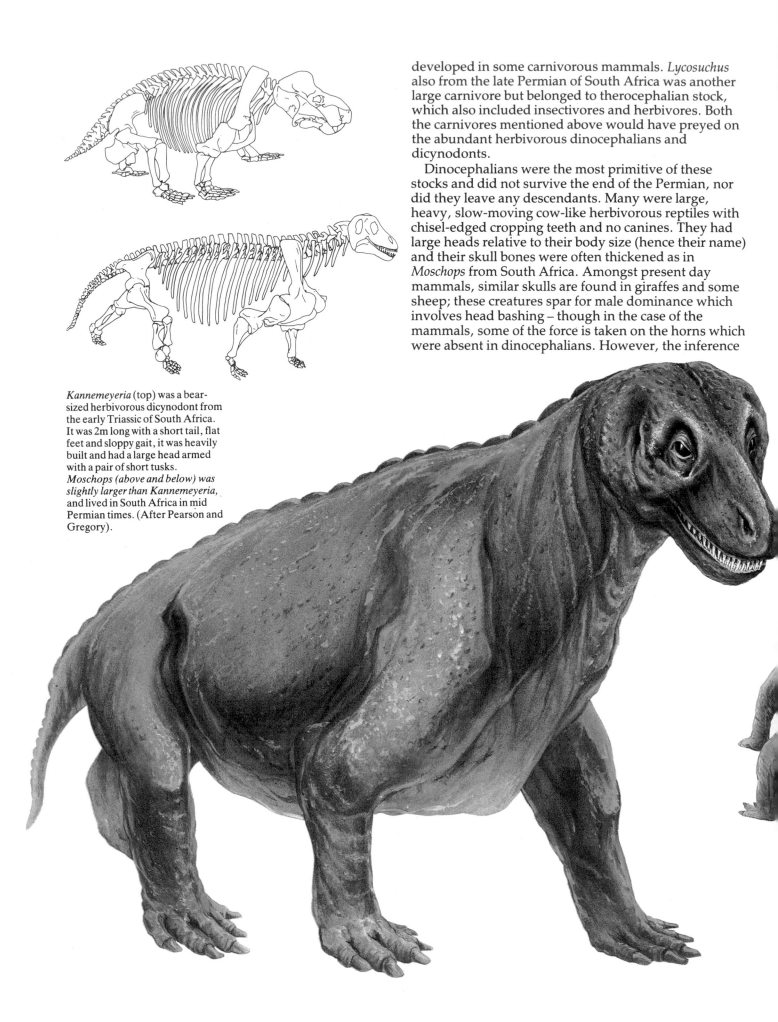

developed in some carnivorous mammals. *Lycosuchus* also from the late Permian of South Africa was another large carnivore but belonged to therocephalian stock, which also included insectivores and herbivores. Both the carnivores mentioned above would have preyed on the abundant herbivorous dinocephalians and dicynodonts.

Dinocephalians were the most primitive of these stocks and did not survive the end of the Permian, nor did they leave any descendants. Many were large, heavy, slow-moving cow-like herbivorous reptiles with chisel-edged cropping teeth and no canines. They had large heads relative to their body size (hence their name) and their skull bones were often thickened as in *Moschops* from South Africa. Amongst present day mammals, similar skulls are found in giraffes and some sheep; these creatures spar for male dominance which involves head bashing – though in the case of the mammals, some of the force is taken on the horns which were absent in dinocephalians. However, the inference

Kannemeyeria (top) was a bear-sized herbivorous dicynodont from the early Triassic of South Africa. It was 2m long with a short tail, flat feet and sloppy gait, it was heavily built and had a large head armed with a pair of short tusks.
Moschops (above and below) was slightly larger than Kannemeyeria, and lived in South Africa in mid Permian times. (After Pearson and Gregory).

seems to be the same, that dinocephalians went in for extensive head bashing. Great numbers of the group have been found in the Karroo Formation of South Africa where these creatures must have roamed in large herds in a climate not unlike the present on the high veldt, and feeding on the abundant bushes of the seed-fern *Glossopteris*.

In the synapsids reviewed so far there is a noticeably wide variety of adaptations with some very active forms. Many had large skulls, well-differentiated dentitions, some with secondary palates and limbs that look

increasingly mammalian. With the cynodonts these trends towards the mammalian condition achieved their goal in mid or late Triassic times. *Thrinaxodon* is a well-known small carnivore from the early Triassic of South Africa with a quite mammalian appearance in its limbs and face, though the backbone and ribs are very unmammalian. *Cynognathus* is another carnivore from the early Triassic of the Karroo in South Africa; it was as its name suggests rather dog-like facially, and it had limited tooth replacement.

From the mid Triassic of Argentina come a series of very advanced cynodonts whose skulls, jaw articulations and dentitions all come very close to being

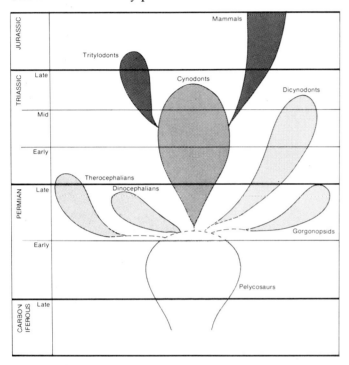

Relationship of synapsid reptiles.

The skull roof of *Moschops* was composed of very thick bone, designed to withstand head bashing contests rather as goats, but without the horns. The anterior teeth intermeshed like a rake and the posterior dentition was much reduced, thus resembling its reptilian cousin, the dinosaur *Diplodocus*.

MAMMAL LIKE REPTILES AND EARLY MAMMALS

Class REPTILIA

Subclass	Synapsida	
Order	Pelycosauria	*Archaeothyris*
		Varanosaurus
		Dimetrodon
		Edaphosaurus
Order	Therapsida	
Suborder	Dinocephalia	*Moschops*
		Titanophoneus
		Ulemosaurus
Suborder	Dicynodontia	*Dicynodon*
		Lystrosaurus
		Kannemeyeria
Suborder	Gorgonopsida	*Scymnognathus*
Suborder	Therocephalia	*Lycosuchus*
Suborder	Cynodontia	*Thrinaxodon*
		Cynognathus
		Probainognathus
Suborder	Tritylodontia	*Oligokyphus*
		Bienotherium

Class MAMMALIA

Subclass	Prototheria	
Order	Triconodonta	*Megazostrodon*
Order	Monotremata	*Ornithorhynchus*
Order	Multituberculata	*Haramiya*
		Taeniolabis
		Ptilodus
Subclass	Theria	
Order	Pantotheria	*Kuehneotherium*
Order	Marsupialia	*Alphadon*
Order	Insectivora	*Zalambdalestes*
		Kennalestes

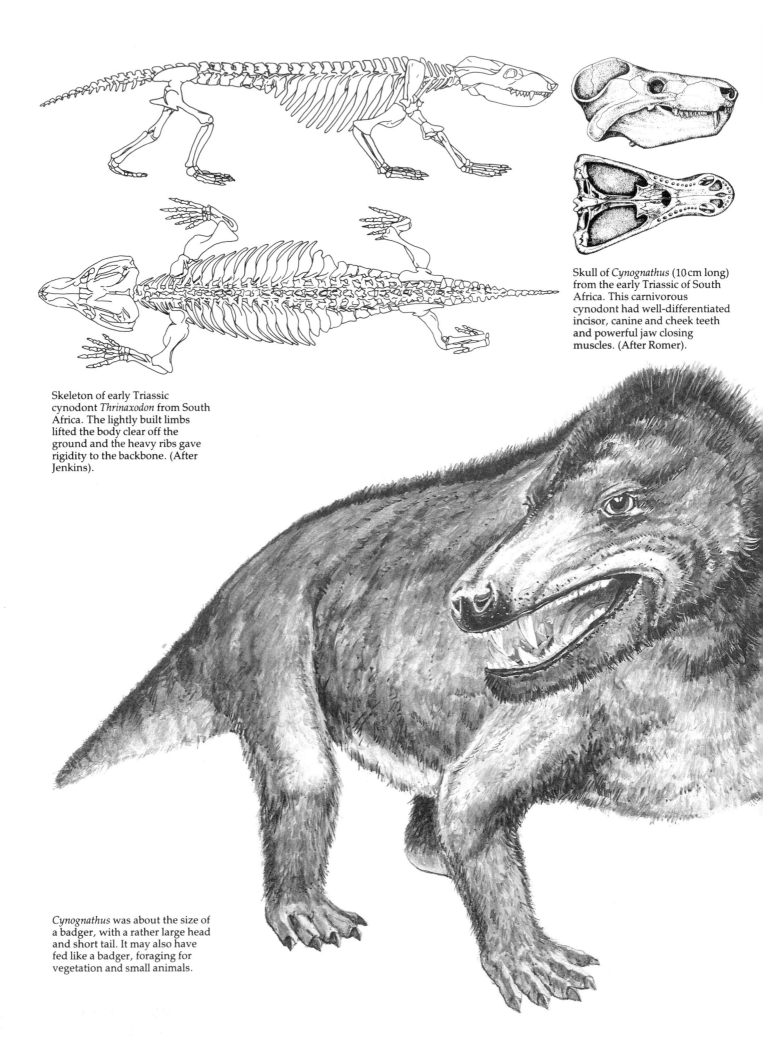

Skeleton of early Triassic cynodont *Thrinaxodon* from South Africa. The lightly built limbs lifted the body clear off the ground and the heavy ribs gave rigidity to the backbone. (After Jenkins).

Skull of *Cynognathus* (10 cm long) from the early Triassic of South Africa. This carnivorous cynodont had well-differentiated incisor, canine and cheek teeth and powerful jaw closing muscles. (After Romer).

Cynognathus was about the size of a badger, with a rather large head and short tail. It may also have fed like a badger, foraging for vegetation and small animals.

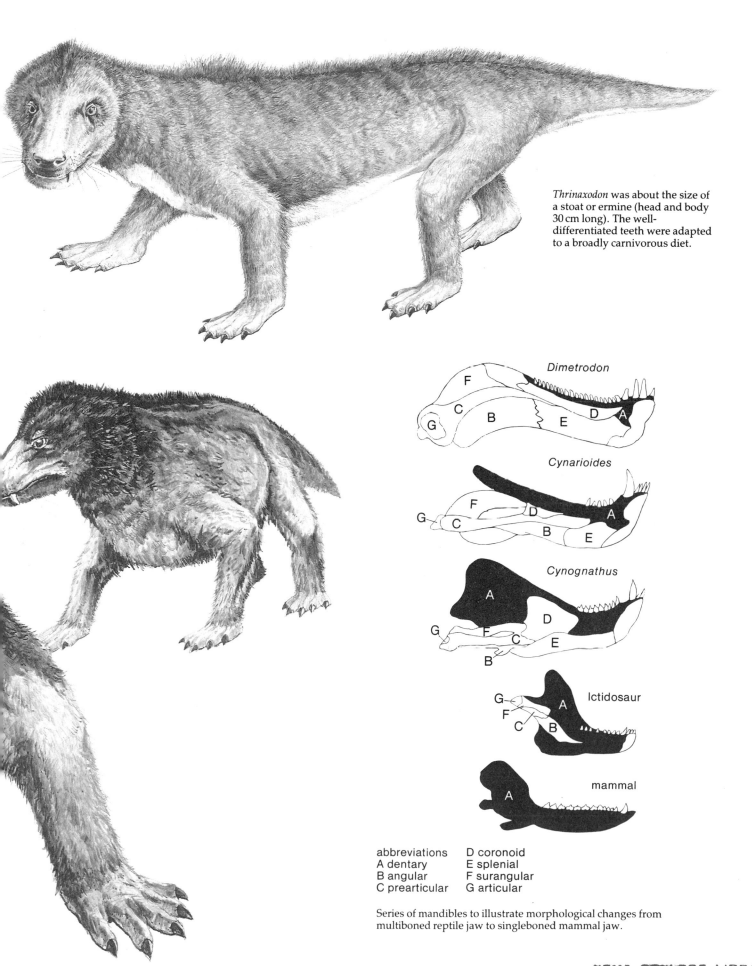

Thrinaxodon was about the size of a stoat or ermine (head and body 30 cm long). The well-differentiated teeth were adapted to a broadly carnivorous diet.

Dimetrodon

Cynarioides

Cynognathus

Ictidosaur

mammal

abbreviations	D coronoid
A dentary	E splenial
B angular	F surangular
C prearticular	G articular

Series of mandibles to illustrate morphological changes from multiboned reptile jaw to singleboned mammal jaw.

mammalian. In *Probainognathus* the jaw articulation was essentially double, possessing both the reptilian (quadrate-articular) and the mammalian (squamosal-dentary). From this it follows that the middle ear had only one bone and so technically it was still a reptile, but in almost every other anatomical feature, it was a mammal. Here then was a creature right on the boundary of our arbitrary division between the two great classes Reptilia and Mammalia. One final group that was important in the story of mammal evolution was the tritylodonts. These animals were an aberrant sideline which persisted long after all the other cynodonts had become extinct. A typical form was *Oligokyphus* from the early Jurassic of England and other forms are known from the Americas, South Africa and China. They were rodent-like animals with a dentition comprising large incisors, which were followed by a gap and then a row of multi-cusped cheek teeth for grinding. They also had the ability to twist their backs and so may have curled up to keep warm and perhaps even suckle their young. Despite this, these little creatures are placed not among the mammals but the reptiles, because their jaw and ear bones were similar to other advanced cynodonts.

By mid Triassic times 230 million years ago, the earliest mammals were emerging on earth. We do not know the exact time or place, but it seems probable that the mammals arose from one cynodont stock and rapidly spread to most continents. As they arose, the other cynodonts and remaining synapsids became extinct, to be replaced by the dinosaurs as the dominant terrestrial vertebrates for the next 150 million years. The mammals had a long wait in the wings, but during that time they did one very important thing – they survived.

Restoration of *Oligokyphus* in an early Jurassic setting of vegetation by a river where they lived rather like present-day water voles. *Oligokyphus* was a very successful vole-like reptile. The tritylodonts outlived all other synapsids, surviving into the mid Jurassic; their remains are known from Eurasia, Africa and America. (After Kühne).

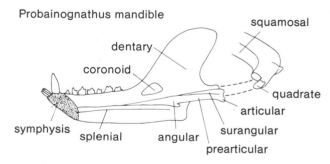

Mandible of *Probainognathus* from the mid Trias of Argentina showing two articulation facets, one reptilian (quadrate/articular) and the other mammalian (squamosal/dentary). (After Romer).

The first true mammals

The earliest known true mammals have come from late Triassic sediments in Europe, China and southern Africa, and already three distinct stocks are evident, which we shall typify with three examples – *Haramiya*, *Megazostrodon* and *Kuehneotherium*. *Haramiya* is known only from the thimblefull of isolated teeth found in Britain and Germany; it was about the size of a mouse and its cheek teeth suggest it crushed food. The second stock is much better known; there is a complete skeleton of *Megazostrodon* from Lesotho in southern Africa, and related genera are recorded from Britain and China. *Megazostrodon* was very shrew-like in size, proportions

 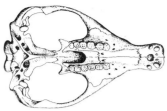

Lateral and ventral views of skulls of, left, *Oligokyphus* (9 cm long) and right, *Bienotherium* (12.5 cm long). Both are early Jurassic tritylodont reptiles which were very rodent-like in appearance and in feeding habits. Note the anterior 'gnawing' teeth and behind the grinding teeth as in voles and agoutis. (After Kühne and Hopson).

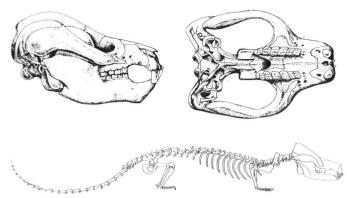

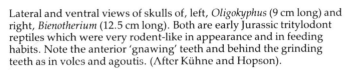

Skeleton of *Oligokyphus* (length 50 cm). (After Kühne).

and mode of life. There is evidence that the mandible closed with an inward movement to provide the shearing action with which the teeth sliced food; the cheek teeth were differentiated into premolars and molars and had limited replacement; the head/neck joint allowed considerable freedom of movement, the structure of the vertebral column allowed much more flexibility than found in any cynodont, the fore limb was well innervated for precise movements, and the hind foot was capable of grasping. The body temperature was probably around 25–30°C, which is rather lower than that found in living mammals, and would be compatible with reptilian metabolic rates. We can envisage

Megazostrodon scrambling over twigs and leaf litter in woods, foraging for insects and grubs during the nocturnal hours. *Kuehneotherium* is the only known genus of the third Triassic mammal stock and known by teeth and bones recovered from fissure infillings in Carboniferous Limestone of Wales; though much less well known than *Megazostrodon*, it may have had a similar mode of life. There were, however, important differences in the way in which the cheek teeth cusps were arranged, the main cusps forming a triangular pattern rather than aligned axially as in *Megazostrodon*; a difference that is of great evolutionary importance in later Mesozoic times.

Restoration and skeletal reconstruction of *Megazostrodon*, one of the earliest mammals from the late Triassic of South Africa. (After Jenkins & Parrington).

The Jurassic and the ancestors of modern mammals appear

Multituberculates, the first herbivorous mammals

The Jurassic mammals can be viewed as specialized and diversified descendants of the Triassic stocks. From the haramiyids evolved the multituberculates, found in the Jurassic of Europe and North America. They were very rodent-like, and it is significant that they replaced the rodent-like tritylodonts between mid and late Jurassic times, and were themselves replaced by true rodents in the late Palaeocene. Jurassic multituberculates are best known from Portugal, where good skull material has been described, though skeletons are still very inadequate. Multituberculates were the first mammalian herbivores, with rodent-like incisor teeth in front of the jaws, then a gap, and a series of gnawing cheek teeth. In multituberculates these teeth possessed many cusps, hence their name; the first lower premolar was often very enlarged as a massive grinding tooth. The brain case reveals that the brain had large frontal lobes and that the area concerned with smell sensing was highly developed.

Triconodonts and docodonts

From the *Megazostrodon* stock evolved the triconodonts and docodonts, again insectivorous mammals and known from Europe and North America. Triconodonts had cheek teeth with three prominent cusps in a row and docodonts had more complex teeth.

Pantotheres, ancestors of modern mammals

Greatest interest centres on the third stock, the pantotheres, a group that evolved from kuehneotheres and included the ancestors of almost all later mammals. The first almost complete skeleton of a pantothere has only recently been found in Portugal; in size and proportions it was similar to a tree shrew (head and body length around 20 cm). The front legs were about 20 per cent shorter than the hind legs, and with its very long tail it seems well adapted for dwelling in trees, jumping from branch to branch, and feeding on insects, grubs and fruit. The triangular arrangement of the main cusps on the cheek teeth, combined with the presence of a marked heel on the lower molars, would have enabled the animal to both slice and crush food, thus giving it a wide dietary potential. Marsupial bones were present in the pelvic region, suggesting it probably had marsupial type reproduction.

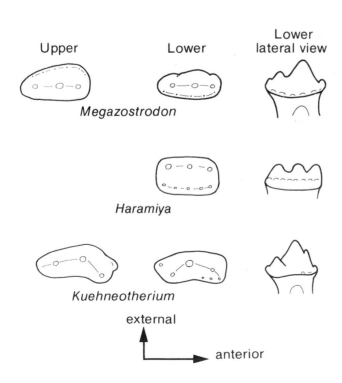

| Upper | Lower | Lower lateral view |

Megazostrodon

Haramiya

Kuehneotherium

external

anterior

Crown and lateral views of cheek teeth of some Mesozoic mammals. Throughout this text the diagrams of teeth are orientated in this way – external to the top and anterior to the right.

Skeleton of a late Jurassic pantothere from Portugal. It resembled a living tree shrew in size and body proportions. The long tail (black) may have been used to hold on to branches. The presence of marsupial bones (stippled) in the pelvic region was probably common to most early mammals. (After Krebs).

The Cretaceous – mammals inherit the earth

Mammals from Mongolia

In Cretaceous times we see the continued success of multituberculates, the disappearance of the triconodonts, and the expansion and diversification of pantothere descendants. As in Jurassic times, we have virtually no evidence about mammals from South America, Africa or Antarctica. However, in the Cretaceous our knowledge from North America and Europe is greatly supplemented by the faunas of Mongolia, which have been collected and studied by Polish palaeontologists; over 150 mammalian skulls are known, mostly around 1.5–2.0 cm long, and of these two thirds are multituberculates and the rest insectivores. The dozen species of multituberculates in the Mongolian Cretaceous testify to the success and diversity of this group at the time.

The origins of marsupials and placentals

From the Jurassic pantotheres we can trace the descent of the two major stocks of living mammals, the marsupials and the placentals, via some as yet poorly known early and mid Cretaceous species from Texas and Mongolia. The late Cretaceous faunas of Mongolia contain eight species of placental mammal; all are essentially insectivorous in their mode of life, though not all are directly related to living insectivores. *Zalambdalestes* is the best known; it was in appearance very like an elephant shrew, and probably walked, ran and jumped similarly using all four legs. From the North American late Cretaceous have come also multituberculates, insectivores and in addition marsupials, which are unknown at any time in Asia.

So with the close of Mesozoic times (65 million years ago) as the dinosaurs declined toward extinction, the mammals were left with the insects, the birds, the lizards, the frogs and toads, and the flowering plants, to usher in the Caenozoic era. Throughout the Mesozoic, mammals had occupied small insectivorous or rodent-like niches; now came their big opportunity – the meek did indeed inherit the earth.

Restoration of *Ptilodus*, a 60-million-year-old North American multituberculate. It probably lived in trees like squirrels.

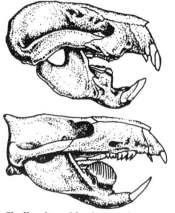

Skulls of two North American Palaeocene multituberculates; *Taeniolabis* (top) and *Ptilodus* (bottom) with view of dentition. Note the rodent-like incisors and the large multicusped grinding cheek teeth. (After Simpson).

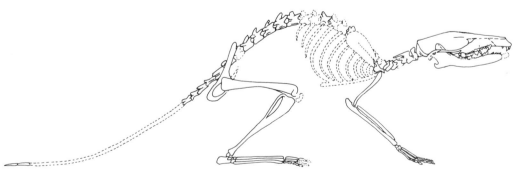

Skeletal reconstruction of *Zalambdalestes* from the late Cretaceous of Mongolia. Head and body about 15 cm long. (After Kielan-Jaworowska). *Zalambdalestes* was similar in size and proportions to modern elephant shrews which live today in the forest undergrowth of tropical Africa.

CHAPTER 5

Insectivores

There are more named species of insect than all other kinds of organisms put together. Insects make up a greater biomass than any other major group of organisms. Collembolans (springtails) have been recorded in Britain with a density of 1.6×10^9 individuals per hectare (=16 per cm^2). The biomass of one large locust swarm has been estimated at half a million tonnes. With such food potential it is not surprising that very many species of land vertebrate – amphibian, reptile, bird and mammal – feed either wholly or partly on insects. Almost all of the known Mesozoic mammals appear to have been insectivorous. Insect feeders are found among 12 orders of living mammals. Just as the order Carnivora does not contain all the carnivorous mammals nor do all members of the order eat meat, so it is with insectivores. The majority of insectivorous mammals today belong to two orders – Insectivora (shrews, hedgehogs and moles) and Chiroptera (bats); in addition there are a few small orders that are mainly insectivorous (tree shrews and elephant shrews); there are also some orders that contain a few species whose diet includes a high proportion of insects (primates and rodents). Finally

The tree shrew, an inhabitant of the forests of south-east Asia. Its generic name *Tupaia* means squirrel in Malay and it resembles a squirrel in some ways, though the face is elongate and pointed. *Tupaia* lives on a mixed diet of insects, fruit, seeds and leaves.

there are the 'anteaters', mammals feeding almost exclusively on colonies of social insects, usually ants and termites.

The anteaters apart, insectivorous mammals are usually small terrestrial mammals, probably little changed from their Mesozoic ancestors. They retain many primitive features and have few specializations. Their origin is often obscure and the fossil record patchy, though nonetheless important. Their classification is difficult and continually changing. The first half of this chapter deals with fossil and living members of the order together with two other small insectivorous orders – Scandentia and Macroscelidea. The second half is devoted to the anteaters and their evolution on different continents. Discussion of bats, rodents, primates and edentates is to be found in later chapters, on the basis that their insectivorousness is secondary to other aspects of their biology.

Order Insectivora

The earliest known placental mammals were insectivores and the niches are filled today by shrews, moles and hedgehogs. These living forms are arranged in seven families, about 60 genera and some 345 species, of which 70 per cent are shrews. The fossil record though patchy is very important and comprises about 150 genera, many in extinct families, ranging from mid Cretaceous times onwards. Insectivores are found throughout the northern hemisphere, in Central America, Africa and south-east Asia. However, there are none in South America (with one exception), Australia or Antarctica, either living or fossil. Almost all insectivores are small mammals; they include the smallest known mammal – the dwarf shrew weighing around 2 g – and the largest living form is the otter shrew with a weight of about 1 kg. Many have spines and in all the senses of smell and hearing are well developed. The ear region often lacks an ossified bulla, the tympanic membrane being attached to a tympanic ring. Vision in insectivores is weak and they are nocturnal or crepuscular. Many shrews can echolocate, using this to find their way around in dense undergrowth in the dark and also to locate prey. Their diet is highly varied; besides insects in all stages of development it can include many other invertebrates and even small vertebrates, together with a small amount of fruit, nuts, seeds and assorted vegetable material. The dentition usually remains fairly complete and the cheek teeth have high pointed cusps. The zygomatic arch is usually weak and often incomplete. The postcranial skeleton is little modified from the generalized primitive mammal condition with pentadactyl limbs and plantigrade locomotion, save in the burrowing forms which have shortened their limbs, the fore limbs being highly modified for powerful digging action. A few insectivores are arboreal or semi-aquatic (natatorial).

Classification of insectivore families is controversial. Many go back to early Tertiary or beyond and few possess many recognizable specialized characters. We shall consider all insectivores as falling within two suborders, the Proteutheria and the Lipotyphla.

Proteutheria, the first placental mammals

The Proteutheria are a poorly defined grouping of about eight families without any living representatives. They exhibit a perplexity of primitive characters, some approaching the primates or other orders, so that it is difficult to draw a dividing line. Some writers would exclude all these stocks from the order Insectivora, but there is no concensus on where else to place them within classifications. Thus they are grouped here among the insectivores more for convenience than from conviction. Most occur in the early Tertiary, but some extend back into the Mesozoic. They are known only from North America and Eurasia, and possibly from North Africa.

INSECTIVOROUS MAMMALS

†*Proteutheria*		
†Endotheriidae		†*Endotherium*
†Pappotheriidae		†*Pappotherium*
†Zalambdolestidae		†*Zalambdolestes*
†Leptictidae		†*Gypsonictis*
†Pantolestidae		†*Buxolestes*
†Ptolemaiidae		†*Ptolemaia*
†Apatemyidae		†*Sinclairella*
†Mixodectidae		†*Mixodectes,*
		†*Elpidophorus*
Lipotyphla		
†Adapisoricidae		†*Adapisorex*
Erinaceidae	Hedgehogs	†*Deinogalerix*
Soricidae	Shrews	*Sorex, Scutisorex,*
		†*Allosorex,* †*Trimylus*
Solenodontidae	Solenodons	*Solenodon*
†Nesophontidae		†*Nesophontes*
†Apternodontidae		†*Apternodus*
†Dimylidae		†*Dimylus,*
		†*Dimyloides*
Talpidae	Moles	*Uropsilus, Desmana*
Chrysochloridae	Golden moles	†*Prochrysochloris*
Tenrecidae	Tenrecs	
Potamogalidae	Otter shrews	
Other insectivorous mammals		
†Cimolesta		
†Palaeoryctidae		†*Cimolestes*
		†*Palaeoryctes*
Macroscelidea	Elephant shrews	†*Metoldobotes,*
		†*Myohyrax,*
		†*Mylomygale*
Scandentia	Tree shrews	*Tupaia*
Marsupialia		
Didelphoidea (pars)	Opossums	†*Garzonia,*
		†*Necrolestes*
Dasyuroidea (pars)	Pouched 'mice'	†*Ankotarinja,* †*Keeuna*
Anteaters		
Monotremata (pars)	Echidnas	*Tachyglossus*
Marsupialia		
Myrmecobiidae	Banded anteater	*Myrmecobius*
Xenarthra		
Dasypodidae (pars)	Armadillos	†*Stegotherium*
Myrmecophagidae	True anteaters	*Tamandua,*
		Myrmecophaga,
		†*Protamandua,*
		†*Eurotamandua*
incertae sedis		†*Ernanodon*
Pholidota	Pangolins	*Manis,* †*Eomanis*
Tubulidentata	Aardvark	*Orycteropus,*
		†*Myorycteropus*
Carnivora		
Hyaenidae (pars)	Aardwolf	*Proteles*

† = Extinct

Restoration of *Zalambdolestes* from the late Cretaceous of Mongolia. In life style and feeding habits this animal probably closely resembled tree shrews.

ENDOTHERIIDS AND PAPPOTHERIIDS A handful of isolated teeth named *Pappotherium* from mid Cretaceous sands of Texas and others named *Endotherium* from an ?early Cretaceous coalfield in Manchuria are probably the earliest placental mammals known. The molar teeth show slight advances on earlier pantotheres from which they appear to be derived. Evidence is, however, much too meagre to include them more than tentatively within the insectivores.

ZALAMBDOLESTIDS The zalambdolestids from the late Cretaceous of Mongolia are well-known from skull and postcranial material, thanks to the painstaking efforts of Professor Kielan-Jaworowska and her Polish and Mongolian colleagues. These little mammals probably looked very like the elephant shrews of today, though not directly related to them. The elongate hind limbs of zalambdolestids as with elephant shrews (and hares) suggest a ricocheting or jumping form of locomotion. The elongate snout would probably have been used as do shrews for locating food by smell and perhaps also by echolocation. Their cheek teeth had the basic placental arrangement of three molars and four premolars; the triangular molar teeth with three cusps, two near the outer margin and the third on the inner apex of the triangle; this basic pattern could be the archetype for most of the later placental mammals. Similarly, the lower molars had a trigonid of cusps anteriorly and a basined talonid behind. The specializations in the dentition of *Zalambdolestes* were the double rooted canine and the procumbent and enlarged lower first incisor tooth.

LEPTICTIDS The leptictids are known from the late Cretaceous of Mongolia, and North America where they persisted until the Oligocene. *Gypsonictis* from the latest Cretaceous of Wyoming has a cheek dentition close to that of zalambdolestids. *Diacodon* from the Palaeocene and early Eocene of the same area is known from a skull

in which the ear region is preserved; this displays a lack of ossified auditory bulla, but the bone arrangement is different from that of primates, thus discounting a common origin for the two stocks. *Leptictis* from the Oligocene was a medium-sized insectivore with a complete dentition $\frac{3143}{3143}$ and no marked specializations. Here, as in *Diacodon*, a hypocone develops behind the protocone on the upper molars – the beginnings of a less triangular and a more squarish molar as seen in later erinaceids. The build of most leptictids and their appearance with an elongate snout, suggest a close similarity to the living gymnures or hairy hedgehogs of south-east Asia.

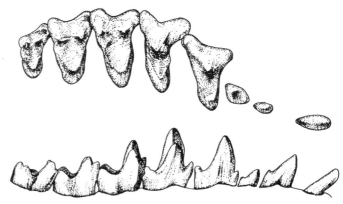

Upper and lower dentition of the late Cretaceous insectivorous mammal *Zalambdolestes*. Note the lower molars have high pointed cusps to pierce soft food and broad crushing shelf behind. (After Kielan-Jaworowska). (Length 15 mm).

The upper cheek teeth of the late Cretaceous leptictid insectivore *Gypsonictis* from Wyoming. (After Lillegraven). (Length 12 mm).

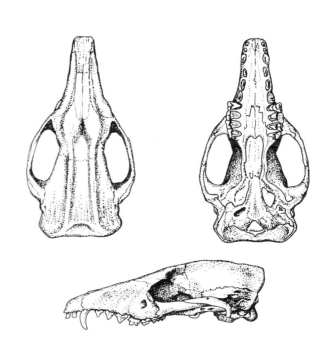

Skull of *Leptictis* from Dakota (length 80 mm). These primitive insectivorous mammals survived into Oligocene times. (After Scott & Jepsen).

Restoration of *Buxolestes* from the Eocene of Germany; the animal was very otter-like in appearance, using its powerful tail and short limbs for efficient swimming.

PANTOLESTIDS Pantolestids were otter-sized animals with semi-aquatic adaptations. Skulls of *Palaeosinopa* (Palaeocene), *Pantolestes* (Eocene) and *Chadronia* (Oligocene) are known from North America, and the complete skeleton of *Buxolestes* has recently been described from Messel (Eocene) in Germany. The molar dentitions of these taxa have broad, thick enamelled and low cusped crowns, usually with an internal cingulum and hypocone; they resemble in appearance the teeth of the sea otter *Enhydra*, and like it they may have eaten molluscs (malacophagous). The tail of *Buxolestes* is very otter like and probably used as the main swimming organ, assisted by the short powerful limbs. Possibly related to the pantolestids is *Ptolemaia* from the Oligocene of Egypt.

APATEMYIDS Apatemyids have some rodent-like features in their anterior dentition, but despite this their skull characteristics are typically insectivore. A small group of medium-sized insectivores they ranged from the Palaeocene to the Oligocene in North America and extended into Europe in the Eocene. Because of their convergences they have been thought of as related to insectivores, primates, rodents and even artiodactyls. The absence of an ossified auditory bulla and the passage of the internal carotid artery in a deep groove in the petrosum allies them with the insectivores rather than with any other stock, but their dentition displays considerable specializations that mark them out as a separate family. In late forms the posterior lower molar develops a very elongate talonid. The first incisor teeth, both upper and lower, become very enlarged; the lower ones are procumbent and chisel edged so that a shearing action occurs when the jaws are closed – not dissimilar to that seen today in the primate *Daubentonia* (aye-aye) and

Top: a lower and two upper molars of *Buxolestes* showing the thick enamelled, broad teeth with low cusps, typical of mollusc feeding mammals. Above: jaws of *Buxolestes* (length 35 mm). (After Jaeger).

Face of an apatemyid, *Sinclairella* displaying the enlarged chisel edged incisor teeth, presumably used to process a special kind of food.

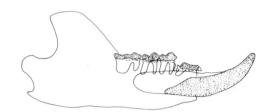

Lower jaw of *Sinclairella* showing the great size of the procumbent chisel edged first incisor tooth. (After Jepsen).

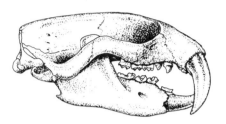 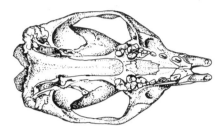 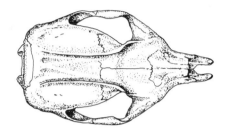

Skull of apatemyid *Sinclairella* from the Oligocene of Dakota. Skull is squirrel-like, but lacks an ossified auditory bulla. (After Scott & Jepsen). (Length 60 mm).

to a lesser extent in some shrews, plesiadapids and microsyopsids; in the case of the apatemyids it was presumably an adaptation to cope with a particular food, but whether that was animal or vegetable is unknown. No postcranial skeletal material exists for apatemyids so that nothing of their mode of life can be deduced. Apatemyids can best be seen as a sideline of insectivore evolution of very limited success.

MIXODECTIDS Mixodectids are a poorly known family comprising three Palaeocene genera in North America and one in Europe. They are known from mandibles and upper jaw fragments and have at various times been allied with primates, rodents, dermoptera and insectivores. They possess enlarged procumbent first lower incisors, the upper canine is double rooted, the anterior premolars are reduced or absent, the posterior premolars are partly molariform, the molars tend to become quadrangular and are low cusped. Until more is discovered their relationships must remain doubtful.

Lipotyphla

The lipotyphla comprise the seven living families of Insectivora, one recently extinct family and four other fossil families. They form a reasonably well-defined group of families with a fossil record mainly in the late Caenozoic. In all known forms the intestinal caecum is lacking and the stapedial artery is the major supplier of blood to the brain.

Mixodectes from the Palaeocene of Wyoming has rodent-like procumbent first incisors, but other dental features separate it from the apatamyids and its relationships are not clearly known. (After Osborn). (Length 60 mm).

ADAPISORICIDS The adapisoricids form a reasonably discrete family from the Palaeocene and Eocene of North America and Europe. They appear to have originated from leptictids and probably include the ancestors of other litophylans. Their dentitions have numerous hedgehog-like (erinaceid) characters; known only from fragmentary remains and isolated teeth it is not possible to build a picture of their form and habit. The molar teeth have high ridged metacone and paracone cusps, with good style and conule development; the hypocone becomes as large as the protocone in some genera, giving the tooth a very squared off appearance as in erinaceids. In the lower dentition the incisors are spatulate and procumbent. The type genus of the family is *Adapisorex*, known from the Palaeocene of France.

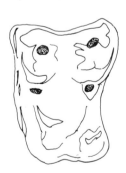

Upper molar of *Adapisorex* from the French Palaeocene; the tooth has hedgehog-like features. (After Russell). (Width 3·7 mm)

ERINACEIDS, THE HEDGEHOGS From an adapisoricid evolved the erinaceids which today comprise two groups – the echinosoricines are the gymnures or hairy hedgehogs of south-east Asia and the erinaceines are the true spiny hedgehogs, found across Eurasia and Africa. Erinaceids are an Old World stock, originating possibly in the Eocene and reaching Africa by the Miocene. Never abundant as fossils, their most spectacular record is that of *Deinogalerix*, the giant hedgehog from a late Miocene fissure infilling at Gargano in southern Italy. This animal was a relative of the hairy hedgehogs and reached the size of a large rabbit. Gigantism is associated also with other members of the fauna, which appears to have been in isolation on an island. The size of

Deinogalerix suggests its diet was not exclusively or even predominantly insects and worms; it probably took flesh, though this may have been scavenged in the absence of means of capturing such prey.

Restoration of the extinct giant hedgehog *Deinogalerix* from the late Miocene of southern Italy. *Deinogalerix* is related to the hairy rather than to the spiny hedgehogs.

Skull and dentition of *Deinogalerix*; note the large incisor teeth probably for seizing prey. (After Freudenthal). (Length 20 cm).

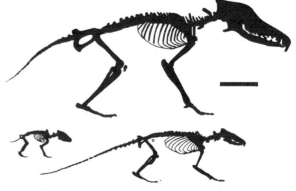

Deinogalerix skeleton to scale with *Echinosorex*, the largest living hairy hedgehog, and with *Erinaceus*, the Old World spiny hedgehog. (After Butler). (Bar is 50 mm).

SORICIDS, THE SHREWS   The soricids or shrews are not only by far the most abundant of living insectivores, with some 245 species in the family, they form the fourth largest family of mammals after two rodent and one bat family. Shrews are insectivores *par excellence*. Many use echolocation and some can secrete a poison in their saliva with which to immobilize or kill their prey. Some have pigmented teeth and pigmentation survives fossilization; it can be traced back to Oligocene shrews. The quadrate upper molar teeth of shrews have an outer W-shaped loph comprising the paracone and metacone; only one lower incisor tooth is present and this is enlarged and procumbent. Fossil taxa are known mainly from North America and Europe, but some have also been found in Asia and Africa; mostly they resemble living shrews.

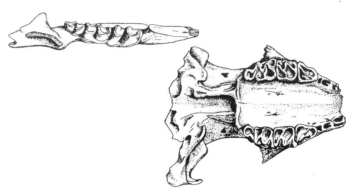

Skull and mandible of the Miocene hedgehog-like shrew *Trimylus* from France. (After Viret & Zapfe). (Length 18 mm).

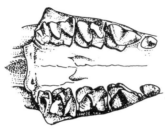

Domnina has typically soricid W-shaped lophs on the molar teeth, (length 10 mm). Oligocene, N. America. (After McDowell).

Upper and lower dentition of the common shrew *Sorex araneus*; note the W-shaped lophs on the molar teeth and the enlarged procumbent lower incisor.

Skeleton of a shrew (*Sorex*) typical of a small primitive insectivorous placental mammal with full set of five toes on each foot.

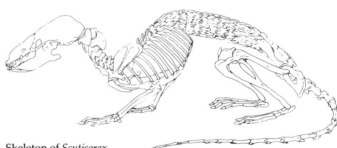

Skeleton of *Scutisorex*.

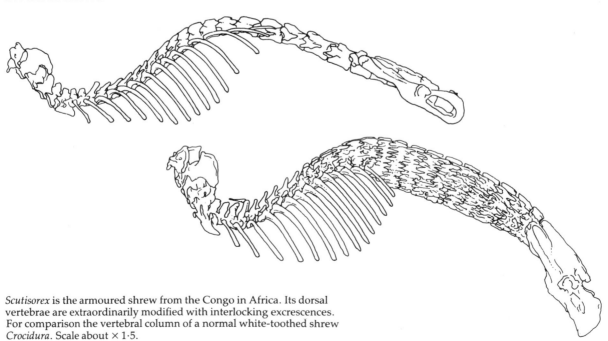

Scutisorex is the armoured shrew from the Congo in Africa. Its dorsal vertebrae are extraordinarily modified with interlocking excrescences. For comparison the vertebral column of a normal white-toothed shrew *Crocidura*. Scale about × 1·5.

Trimylus was one of an extinct subfamily of large hedgehog-like shrews occurring in Oligocene and Miocene deposits of North America and Europe. *Allosorex* from the European Pliocene was probably carnivorous and weasel-like. The strangest shrew is *Scutisorex* living in West and Central Africa; in many ways it resembles *Crocidura*, a white-toothed shrew, but its vertebrae are unique. The lumbar and posterior thoracic vertebrae possess bony excrescences that interlock with adjoining vertebrae. The function is not fully understood, but it is thought to produce a very strong backbone which may enable the animal to force its way under stones in search of food.

SOLENODONTIDS AND NESOPHONTIDS   *Solenodon* and the recently extinct *Nesophontes* are two West Indies shrew-like animals from the Caribbean islands. They have probably been in isolation there since early Caenozoic times for these animals do not have close relationships with any living insectivore and are usually each placed in separate families. *Apternodus* from the Oligocene of North America bears some resemblance to them but may not be ancestral; like the living *Solenodon*, it had triangular upper molar teeth with V-shaped cusps and prominent outer styles.

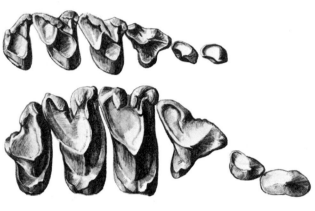

Top: upper dentition of *Nesophontes* (length 13 mm). Above: upper dentition of the West Indian endemic shrew, *Solenodon* (length 17 mm) still living on Cuba and Hispaniola. (After McDowell).

Skull of *Apternodus*, a solenodon-like insectivore from the Oligocene of North America. (After McDowell). (Length 45 mm).

Restoration of *Nesophontes*, an endemic shrew from Cuba and Puerto Rico which probably became extinct in the 1930s. Of six known species, the largest was about the size of a chipmunk.

Restoration of *Dimylus* from the early Miocene of Europe; a desman-like insectivore, taking a mollusc.

DIMYLIDS   Another interesting sideline of evolution is the family Dimylidae; members of this group are known only from the Oligocene and Miocene of Europe by a handful of genera, most represented only by dental fragments. The discovery of a snout with mandible attached has helped us to understand the relationship of these strange beasts. They possess characteristics of both soricids and talpids (moles) and are rather closer to the latter, although their specialization deserves separation at family level. Their teeth have thick enamel, the upper canine is double rooted, the cusps of the cheek teeth are low and the premolars broad and strong – all characters associated with the mollusc feeding mammals such as the sea otters.

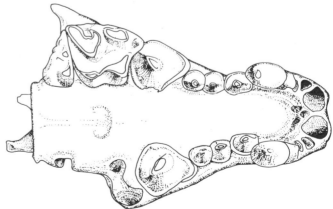

Skull of *Dimyloides* from the Oligocene of Germany. Note the very enlarged cheek teeth with thickened enamelled cusps. (After Schmidt-Kittler). (Length 13 mm).

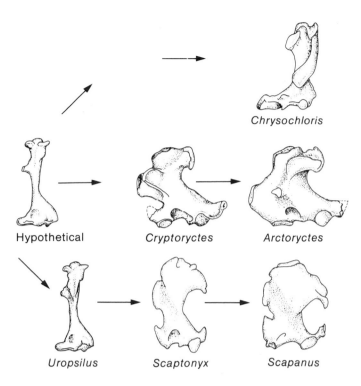

Series of humeri of moles showing three lineages (left to right) in which the width of the humerus greatly increases, thus improving the digging power of these animals. (After Reed & Turnbull).

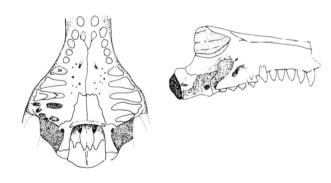

Skull of the Miocene *Prochrysochloris* from East Africa; an early African mole of the family Chrysochloridae. (After Butler & Hopwood). (Length 9mm).

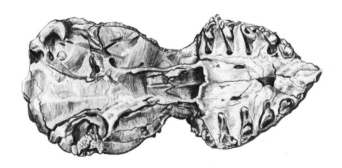

Skull of *Palaeoryctes* from the Palaeocene of southern USA. This, like *Cimolestes*, was a shrew-like mammal close to the ancestry of carnivores. (After McDowell). (Length 11mm).

TALPIDS, THE MOLES The most distinctive features of the talpids is their adaptation for tunnelling. The fore limb and in particular the humerus is in almost all cases highly specialized for earth moving – short and broad with powerful retractor muscles; the bone is very diagnostic when found fossilized. One living form, *Uropsilus* the shrew mole, marks a halfway stage between shrews and moles. The fossil record of the talpids extends back to the late Eocene in Europe and nearly as far back in North America. We do not have evidence of moles in Asia before the late Miocene and they never reached Africa. One subfamily, the Desmaninae, is semi-aquatic; in desmans the hind limb is adapted to a powerful back stroke and used to shift water rather than earth and thus propel the animal forward. A desman lived in Britain during the mid Pleistocene but it is now only found in the Pyrenees, with another species in southern Russia.

CHRYSOCHLORIDS, THE GOLDEN MOLES Another family of moles is the Chrysochloridae or golden moles of Africa, where they fill the niche occupied by talpids on other continents. The fossil record of chrysochlorids is poor; there is a species recorded from the early Miocene of Kenya and three others from the Plio-Pleistocene of South Africa.

TENRECIDS, THE TENRECS On the island of Madagascar is found only one group of insectivores – the tenrecs – with about 10 genera and 30 species. They have diversified to fill niches occupied elsewhere by shrews and hedgehogs; some are spiny, some semi-aquatic and some burrowing (fossorial). An isolated family, their origin can be traced to Africa where in the early Miocene of Kenya there were three forms with close similarities to the living Malagasy species. In Miocene times the tenrecs must have been abundant and varied in Africa, and chance dispersal(s) enabled colonies to become established in Madagascar. While the tenrecs no longer live in Africa, they have very close relatives in the otter shrews (potamogalids) of Central Africa; these are the largest living insectivores and very like small otters in life habit, feeding on fish and crustaceans.

Other insectivorous mammals

Apart from the insect-feeding bats, rodents, primates and edentates dealt with later, there remain four orders of mammals which are wholly or in part adapted to an insectivorous diet.

PALAEORYCTIDS The oldest of these are members of the family Palaeoryctidae (order Cimolesta) known from the late Cretaceous and Palaeocene of North America, Mongolia, Morocco and possibly France. *Cimolestes* in the late Cretaceous was a contemporary of *Leptictis*, but the basic plan of the molar teeth was quite different. As we have seen lepticids appear to be close to the ancestors of many other insectivorous mammals, while the palaeoryctids seem close to the ancestry of the carnivorous order Creodonta. A well-preserved skull of *Palaeoryctes* from the Palaeocene of New Mexico shows it to have been shrew-like in appearance.

Restoration of *Cimolestes* which in life might have resembled an elephant shrew.

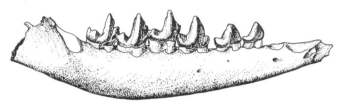

Mandible of *Cimolestes* from the late Cretaceous. *Cimolestes* was an insectivorous feeder with sharp high pointed cusps on its molar teeth and anterior shearing edges on the trigonids. (After Lillegraven). (Length 30 mm).

Lower dentition of *Metoldobotes* from the early Oligocene of Egypt, one of the earliest known elephant shrews. (After Patterson). (Length 30 mm).

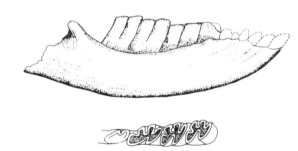

Rodent-like dentition seen in the mandible of *Mylomygale*, a Pliocene macroscelid from South Africa. (After Broom). ($M_1 - M_3 = 10$ mm).

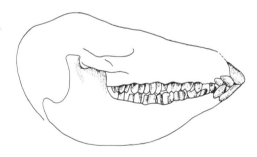

Skull of *Myohyrax*, a hyrax-like macroscelid from the Miocene of East Africa. (After Whitworth). (Length 40 mm).

MACROSCELIDS, THE ELEPHANT SHREWS The elephant shrews differ so much from other insectivores that they are nowadays placed in a separate order – the Macroscelidea. Both fossil and living members of the order are restricted to Africa. They are rather shrew-like but often larger; they have big eyes and are diurnal; their auditory bulla is ossified; they use their hind legs to hop around and feed on insects, fruit, and vegetation shoots; they have a long narrow snout with a proboscis, emit high pitched squeaks and may echolocate. They have a fossil record going back to the early Oligocene with *Metoldobotes* from the Fayum in Egypt; five fossil and four living genera are known. Their cheek teeth are quadrate and ungulate-like, sometimes hypsodont; M_3 is reduced or absent. Some fossil forms were quite rodent-like (e.g. *Mylomygale* from the Pliocene of South Africa) while others were hyracoid-like (e.g. *Myohyrax* from the Miocene of East Africa). These similarities to other groups of mammal have caused the macroscelids to be misidentified or allied with insectivores, hyraxes, rodents, primates, tree shrews and even hares. Most probably the early forms were omnivorous and from these some herbivorous forms developed.

SCANDENTIA, THE TREE SHREWS The tree shrews are another small isolated mammalian order, comprising five genera and about 16 species living in the forests of south-east Asia. Their anatomy has often been interpreted as primitively primate, but nowadays they are placed in their own order, the Scandentia. These squirrel-like mammals have a diet which consists primarily of insects and fruit. They have no fossil record unless the poorly known *Adapisoriculus* from the late Palaeocene of France is a tupaiid.

MARSUPIAL INSECTIVORES The absence of Insectivora from Australia and South America merits attention. Australia until Plio-Pleistocene times possessed only two groups of land mammal – monotremes and marsupials. We are almost totally ignorant of the origins of the monotremes in Australia; the marsupials seem likely to have originated in South America and reached Australia via Antarctica in early Tertiary times. Until Oligocene times South America possessed only three mammalian stocks – condylarths and their ungulate descendants, edentates and marsupials. Even today, despite their abundance in North and Central America, only one genus of shrew (*Cryptotis*) has established itself on the continent in Columbia, Venezuela and Ecuador. Their place has been taken by armadillos and marsupials. Throughout the Tertiary in South America there appear to have been a number of lineages of small marsupials for which insects would have been an important dietary requirement. Today they include *Marmosa*, *Dromiciops* and *Caenolestes*. In the Tertiary at least four groups of marsupials had insectivorous members – didelphids, microbiotheriids, caenolestids and necrolestids. From the early Miocene deposits of Santa Cruz in Argentina, come *Garzonia*, a shrew-like marsupial with a procumbent lower incisor tooth, and *Necrolestes*, a curious creature with an upturned snout and features which resemble a chrysochlorid mole.

In Australia today the small dasyurid marsupials such as *Antechinus* and *Sminthopsis* feed largely on insects. From the Miocene of South Australia come records of the earliest dasyurids, *Ankotarinja* and *Keeuna*, which were probably in part insectivorous.

Mandible of *Garzonia*, an insectivorous marsupial from the Miocene of Patagonia. (After Sinclair). (Length 21 mm).

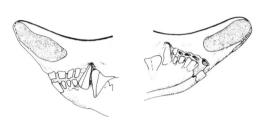

Snout of *Necrolestes*, a unique presumed insectivorous marsupial from the Miocene of Patagonia; no other part of the animal is known. (After Patterson). (Scale × 1·5).

Restoration of *Necrolestes*; the name means 'grave-robber', a strange name for a strange beast with elongate and upturned bony nasal process.

Anteaters

When a shrew captures an insect, it acquires a very small package of high quality protein, but at a high expense in terms of the energy output required to track down and take the prey; the problem is to obtain more energy than is expended and this is a factor which keeps insectivorous mammals small. When insects, however, become social and form large colonies of workers, then the situation is dramatically changed – a concentrated source of food is available with little effort.

The major groups of social insects are ants, termites, bees and wasps. The latter two make the seizure task particularly difficult; they have poisonous stings which few mammals can withstand. *Tamandua*, the collared anteater of Central and South America is one of the few that regularly includes bees in its diet.

Ants are the most abundant social insects. The first known ant is *Sphecomyrma* from the mid Cretaceous of New Jersey, USA and by that time it seems probable that, as it is a worker, ants were fully social. Today we recognise about 7600 species of ant and there may well be double that number unnamed. It has been calculated that at any one instant there are not less than 10^{15} living ants on earth; their total biomass about a million tonnes. Ants are found from the Arctic to Tierra del Fuego and on most oceanic islands; they have very varied and often highly specialized diets. One colony of safari ants can number over 20 million individuals.

The second great group of social insects are the termites or white ants; they are limited to the tropics with around 2200 known species. They are also recorded as mid Cretaceous fossils with *Cretatermes* from Labrador. Termites subsist on a diet of cellulose and some species build earth mounds exceeding 12 metres in height.

The food resources of ants and termites appear almost limitless, but for the mammal there are problems. Other animals have adopted the same strategy – amphibians, reptiles, birds, arthropods and other insects. Also some specialized equipment is needed; a good sense of hearing and smell to locate the colonies, strong front limbs with powerful claws to break open the nests, a long sticky tongue with which to catch the insects, a long hard palate to prevent ants going down the windpipe, a digestion to cope with formic acid and a resistance to ant and termite bites. While acquiring these features others

A worker ant *Sphecomyrma* preserved in amber of mid Cretaceous age in New Jersey, USA. This is the first evidence of social ants with division of labour into different types of individual, which made large ant colonies possible. (After Wilson).

can be lost or reduced. Teeth become largely superfluous and are shed; mastication musculature can likewise be dispensed with; the zygomatic arch can be abandoned and the mandibular articulation can be loose. Given all this, success is within reach and seven groups of mammal have independently taken the option. They are distributed on all continents except Antarctica. The adaptation to a common diet has resulted in strong convergences in the above characters, but there are many differences. Their size varies from the equivalent of hedgehog to that of pig. Their pelage may be long or short hair, almost hairless with thick hide, spines, horny shields or scales. Most have reduced their dentition and many are toothless; some are arboreal and some terrestrial; most can burrow rapidly.

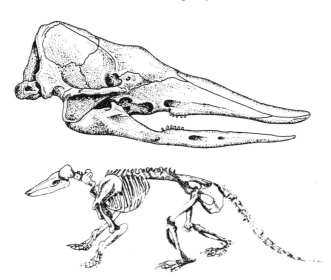

Skull and skeleton of the armadillo *Stegotherium* from the early Miocene of Patagonia; already it shows clear specializations to ant feeding in its long almost toothless jaws. (After Scott). (Skull 14 cm long).

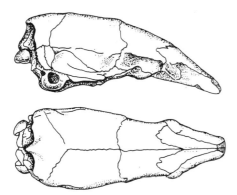

Palaeomyrmedon is a Pliocene myrmecophagid from Patagonia. (After Rovereto). (Skull 10 cm long).

In Australia termite mounds can reach 6 m in height and have a base diameter of 30 m. These vast structures may weigh thousands of tonnes and in deserts the termites may bore a shaft down to 40 m to reach water. They occur abundantly in the tropics of all southern continents.

Monotreme and marsupial anteaters

Australasia has echidnas and numbats. The echidnas of Australia and Papua New Guinea are hedgehog-like spiny anteaters belonging to the order Monotremata – egg laying toothless mammals. Although so primitive in many ways, they have a very poor fossil record; the earliest is the platypus *Obdurodon* from the Miocene of South Australia. The numbats or banded anteaters are marsupial and they specialize in digging out termite nests; captive individuals can eat 20 000 termites a day (equivalent to about 15 per cent body weight). Their fossil record only goes back to the Pleistocene, but analysis of their blood indicates that they are a very ancient group separating from the dasyurids in early Tertiary times.

The living Australian marsupial anteater *Myrmecobius*. Found in mulga scrub in southern Australia, it feeds mainly on termites.

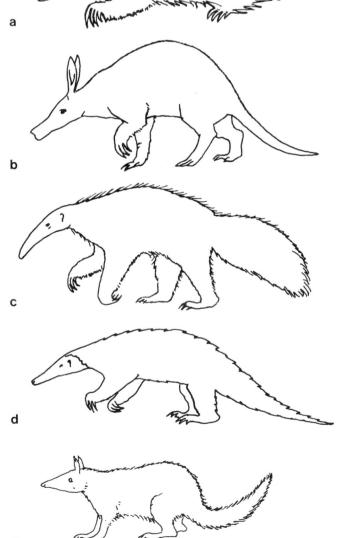

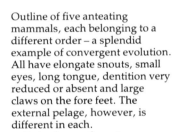

Outline of five anteating mammals, each belonging to a different order – a splendid example of convergent evolution. All have elongate snouts, small eyes, long tongue, dentition very reduced or absent and large claws on the fore feet. The external pelage, however, is different in each.
a. *Tachyglossus*, the spiny anteater from Australasia; a monotreme with spines.

b. *Orycteropus*, the aardvark from Africa; a tubulidentate with naked skin.
c. *Myrmecophaga*, the giant anteater from South America; an edentate with very long hair.
d. *Manis*, the scaly pangolin from Africa and Asia; a pholidote with scales.
e. *Myrmecobius*, the pouched anteater from Australia; a marsupial with short hair.

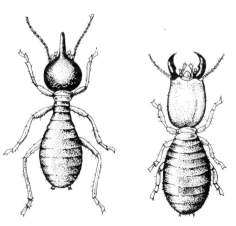
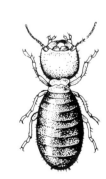

The diet of anteaters. The two termites on the left are soldiers of the species *Nasutitermes* and *Amitermes*, the one on the right is a worker.

Xenarthrans, armadillos and true anteaters

In South America marsupial anteaters do not appear to have evolved but their place is taken by two stocks of xenarthrans – dasypodids (armadillos) and myrmecophagids (true anteaters). About half a dozen species of living armadillo of Central and South America have a diet composed largely of termites and ants. They have an armour of horny plates arranged in bands over the back to allow mobility, with shields over the head, and around the tail and limbs. They have very powerful fore limbs and are terrestrial. *Priodontes* the giant armadillo can reach 60 kg. The earliest fossil armadillo occurs in the late Palaeocene of Argentina, though at this stage it may not have been anteating (myrmecophagous). By the Santacrucian (early Miocene) armadillos were varied and abundant; *Stegotherium* has a long toothless snout and was clearly anteating.

The myrmecophagids live in tropical forests and savannah. There are only three genera – the giant, the collared and the two toed anteater. They have long tapered toothless snouts and very long tongues, small ears, a very long hairy tail (prehensile in the case of the climbing *Cyclopes*). The fore limbs are strong with very robust claws. The giant anteater *Myrmecophaga* can weigh 27 kg. The fossil record in South America is poor, but the first known specimens in the Santacrucian named *Protamandua*, were toothless and obviously anteaters. Until very recently these myrmecophagids were believed to be entirely restricted to the South American continent but now a complete skeleton has been found at the mid Eocene site of Messel in Germany; it is named *Eurotamandua* and is very similar to *Tamandua*, the collared anteater. The site also preserves fossil ants!

Eurotamandua from the brown coal deposit of Messel in Germany. This mid Eocene anteater is very closely similar to *Tamandua*, the living collared anteater of South America; both belong to the same family, the Myrmecophagidae. (After Storch). (One third natural size).

Restoration of *Eurotamandua* with young, licking termites from a mound.

Pholidotes, pangolins

Europe, Asia and Africa have records of another two myrmecophagous mammalian orders, the Pholidota (pangolins) and the Tubulidentata (aardvarks). The pholidotes are known from the living genus *Manis* with seven species in tropical Africa and south-east Asia. They include terrestrial and arboreal species and all have the usual anteating features of long toothless snout, long tongue, long tail, good sense of smell, strong fore limbs and large claws. Pangolins are covered in

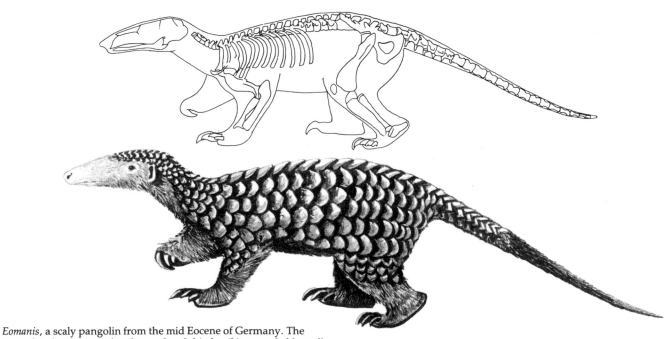

Eomanis, a scaly pangolin from the mid Eocene of Germany. The pangolins have a poor fossil record and this fossil is remarkably well preserved, even to having its scales intact as seen in the living *Manis* from Africa and south-east Asia. (After Storch). (Total length 50 cm).

scales, making them look superficially like pine cones; they can close eyes, nostrils and ears to keep ants out. *Eomanis*, the earliest known pangolin is preserved at Messel alongside the *Eurotamandua*; the stomach contents of *Eomanis* are intact and include vegetable matter as well as insects. Limb bones characteristic of pangolins are known from the Tertiary of North America, Western Europe, India and East Africa.

A recent discovery has been announced of a strange xenarthran-like animal in the late Palaeocene of China. The animal has been named *Ernanodon* and combines features found in sloths, pangolins, aardwolf, aardvark and pandas; it is thought to represent an early and independent lineage specializing in ant or termite eating.

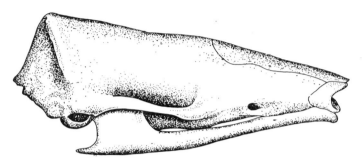

Skull of *Eurotamandua*; the elongate snout, weak mandible and toothless jaws closely resemble the living anteaters. (After Storch). (Length 10 cm).

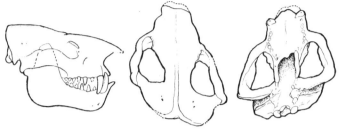

Skull of *Ernanodon*, the enigmatic anteater from the Palaeocene of China. (After Radinsky & Ting). (Skull length 12 cm).

Tubulidentates, the aardvark

The only living tubulidentate is *Orycteropus*, the aardvark (meaning earthpig in Afrikaans). It is pig-like with a long snout and unusually for anteaters has big ears. It is an efficient burrower and the largest of anteaters, reaching 70 kg. The fossil record is patchy – mostly limb bones and jaw fragments which retain enamel-less peg-like teeth. The earliest known is *Myorycteropus* from the early Miocene of East Africa. Other fossils are known from the Tethyan seaboard (France, Greece, Turkey, India) and from the Pleistocene of Madagascar.

While the dasypodids and myrmecophagids are clearly related families of the order Xenarthra, the relationship of this order to the Pholidota and the Tublidentata is not understood. Living taxa do not suffice to unravel the problem and the fossil record is sparse.

Hyaenids, the aardwolf

Finally we glimpse a quite different anteater, the aardwolf *Proteles*. A hyaenid with very reduced dentition of short single cusped teeth, the living aardwolf of Africa has no fossil record. However, it seems eminently reasonable that a carnivore should adapt to anteating – which is in essence a specialized form of carnivory.

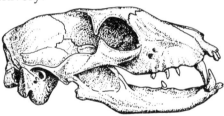

Skull of *Proteles* showing the elongate snout, weak jaws and vestigial teeth. (Skull length 14 cm).

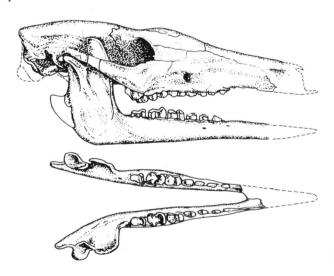

Skull and mandible of *Orycteropus gaudryi*, a late Miocene aardvark from the island of Samos in the Mediterranean; it is similar to but smaller than the living aardvark. (After Colbert). (Skull length 18 cm).

Proteles, the aardwolf of Africa; this strange termite feeder is a close relative of the hyaenas.

CHAPTER 6
Carnivores

The lion is widely regarded as the king of the beasts, yet to most of us he is most familiar as a large lazy animal dozing under a tree in a game park. To find out if he justifies his title let us see how he sets about earning his living, how he catches his supper. The first and most important lesson is spot your supper before your victim spots you; this requires keen sensory perception, particularly vision, and the ability to process the information efficiently in the brain. The next step is to stalk up close to your victim, taking full advantage of wind and vegetation cover; this may involve crawling along, belly-scraping the ground. When the distance has been shortened to not more than 200 metres, and preferably less, the exact moment must be chosen for the final charge. Lions are lightly built and do not have the stamina for a long race, they are short distance sprinters. The impact of lion and victim, usually a zebra or antelope, will take place between 50 and 60 kh^{-1}. Two things should happen; the victim be thrown off balance and the lion's canine teeth penetrate the victim's neck, puncturing vital blood vessels. While the lion's teeth penetrate the neck, the front claws are used to hold on to the victim's neck and the hind claws to tear the belly; in this position the lion escapes fierce kicking which could severely injure him. If all is successful, the lion can soon start using his specialized carnassial teeth to slice up the meat into swallowable chunks, provided he is also successful in warding off would-be thieves such as hyaenas or hunting dogs. With all the refinements of millions of years of evolution, the lion is only successful about one in five times; it is clearly not a game for amateurs.

The two most essential characteristics which define a carnivore are (1) a diet which comprises a substantial proportion of vertebrates, be it fish, reptile or mammal, and (2) at least one pair of specialized shearing or carnassial teeth with which to slice the meat. It is impossible to be more precise; the same species may have different diets in different places and at different times of year. In fossil species the diet can only be guessed. In many the diet is very varied with an overlap into insectivorous and herbivorous types. The marine carnivores feeding on fish have lost their carnassial teeth; as also have the mollusc crushers. Wild pigs may kill and even eat the occasional small mammal, but they would hardly rank as carnivores. The whales are excluded from this section because they are primarily adapted to life at sea and it is thus as aquatic mammals that they are considered. The carnivorous mode of life is found among three major mammalian orders – a few families of marsupials, all of the extinct creodonts, and most of the families of living carnivores. Hence the taxonomic order Carnivora is not synonymous with carnivorous mammals, and several members of the order do not eat meat; such are the panda (a bamboo eater), the badger (an omnivore) and the aardwolf (an ant and termite eater).

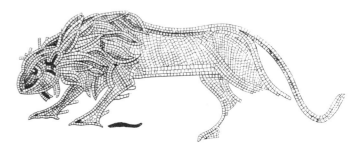

Lion from floor mosaic of a Roman villa in Gloucestershire, England. The lion is one of a series on animals in the Orpheus mosaic. Second century AD

Carnivore characteristics

Size

Present-day carnivores make up about three per cent of the mammalian biomass. The majority of them are medium-sized animals (averaging around 4 kg) with rodents as their main diet. Without these stoats, cats, dogs, and civets the world would soon be overrun by rats and their kind. The smallest carnivore is the least weasel which weights about 50 g, or about ten times as much as the smallest mammal, a shrew. The largest living true carnivore is a tiger, weighing up to 275 kg. Bears can reach 800 kg, but are omnivores rather than true carnivores. Even a bear is very much smaller than the largest living land mammal, an elephant which may weigh some 6 tonnes. Some fossil elephants probably weighed around 8 tonnes, which would be around ten times the weight of the largest carnivore. A carnivore which approached an elephant in size would not be able to run fast and so would have great problems in catching its food. Also it would require a very large supply of these gigantic beasts, for the smaller ones would all be able to outpace it. To overcome this problem the more highly evolved carnivores hunt in groups or packs, several individuals combining to bring down a victim.

Anatomy and physiology

Compared with herbivores, carnivores have relatively few anatomical and physiological specializations. The

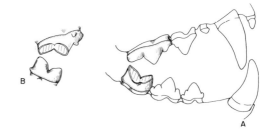

A cat jaw to show the scissor-like action of the carnassial teeth. (a) the jaws viewed from the outside; (b) carnassial teeth viewed from the inside. The cutting edge is on the inner side of the upper carnassial and on the outer side of the lower carnassial.

main features of their skeleton relate to the need for versatility and for strength. A carnivore may have in the course of living to walk, run, climb, jump, crawl, pounce, grasp, tear, dig and swim. Its limbs and backbone must be strong, flexible and multipurpose. The tail is often long and used in balance, in climbing and in retaining heat when wrapped round the body in sleep. The limbs are usually of near equal length with the hind limbs about ten per cent longer than the front (except that in hyaenas the reverse is true). In aquatic and fossorial forms the limbs are short, in fast running forms they are elongate. The feet usually have four or five claws; bears, raccoons and primitive fossil forms are flat footed (plantigrade, like men). In cats, dogs and other fast running forms, the feet are more lightly built and only the toes touch the ground (digitigrade as in women with high heeled shoes, except that they do not seem to be able to run very fast). Most cats and a few other forms have retractile claws, which enables them to keep their claws sharp for use in battle.

SENSES   The shape of the skull is essentially the product of the requirements of the senses and the feeding apparatus. In carnivores the brain is usually relatively large, and the complexity of the convolutions indicates a high degree of control of motor functions and efficient processing of sensory information. Eyes in modern carnivores give stereoscopic vision of high resolution – a cheetah making contact with a victim at 100 kh^{-1} has to lodge a canine tooth 1 cm diameter into a blood vessel of similar width; this is very high precision work. The ears of carnivores have a greater frequency range than in man, and in animals like the desert fox (*Fennecus*) and the bateared fox (*Otocyon*) are the main sensory organ. The olfactory sense is usually well developed, and especially so in fossorial forms.

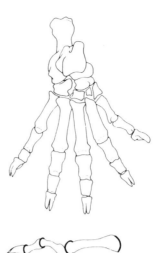
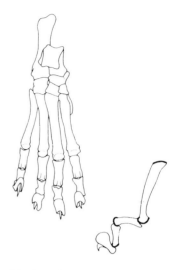

The hind foot of an extinct plantigrade creodont carnivore with short broad digits, well spread out; metatarsals and phalanges in contact with the ground.

The hind foot of a living cat with long thin digits, held close together; only the phalanges touch the ground, hence the term digitigrade is used to describe them.

Illustration of the retractile claw of a cat. By contracting a muscle on the lower surface, the claw is extended for use; by contracting a muscle on the upper surface, the claw is retracted.

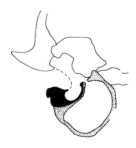

Section through the ear of a cat. The large auditory bulla is composed of two tympanic bones (black and stippled), with a shelf dividing the space into two chambers. The ear drum is represented by the dashed line. The bulla serves as a resonance chamber for sound on its way to the inner ear.

The bateared fox *Otocyon* lives in arid parts of southern and eastern Africa and feeds on a mixed diet of insects (including termites), fruit, tubers, and small lizards and rodents. As the large ears suggest, it has very acute auditory perception.

DENTITION In carnivores, as opposed to herbivores, the jaws have a cylindrical hinge which is level with the dentition and the temporal muscle is the main power muscle used in closing the jaw. The most powerful muscles are found among the bone crushers, which comprise the hyaenas of today and some dogs and creodonts in Tertiary times. The dentition in carnivorous mammals is usually not very different from a primitive placental mammal like a shrew; it could be argued on dentition that a stoat is simply an overgrown shrew. Almost all carnivorous mammals retain three pairs of incisor, used in tearing, grasping and grooming. The canine teeth are always well developed, and in some groups extremely developed into sabre teeth for stabbing. The premolars are often reduced but the posterior premolars become enlarged, broad and heavily enamelled in the bone crushers. The major dental characteristic of carnivores is the specialization in the molar dentition, and that can best be understood by tracing the evolution from its Mesozoic ancestors.

We have seen how in *Cimolestes* the molar dentition has a twofold function, to slice and to crush. In carnivores it is the slicing function that is emphasized, often at the expense of crushing. A dog uses P^4 and M_1 to slice food and the teeth behind these are flattened for crushing. In a cat the same pair of teeth are used for slicing, but there are no teeth behind for crushing; it just cuts and swallows, it cannot chew its food. As described below there are many differing combinations that have been utilized by carnivorous mammals. The slicing or carnassial teeth act on meat rather like a pair of scissors; the blade of the upper and lower tooth are brought past each other by the powerful closing muscles of the jaw, and are kept in close contact by the cylindrical hinge which allows only up and down movements. A pair of scissors with a loose hinge allows the blades to separate and it will not then cut; the same principal holds when a cat slices meat. Further the scissors only work if their blades are sharp. In carnivores, by acting against each other, the blades are self sharpening.

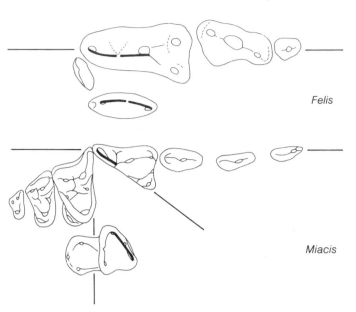

Felis

Miacis

The blade of the carnassial shearing tooth in early carnivores such as *Miacis* (above) is short and diagonal to the dental row. In advanced carnivores such as the cat *Felis* the blade is elongate and parallel to the dental row and there are not crushing molars behind.

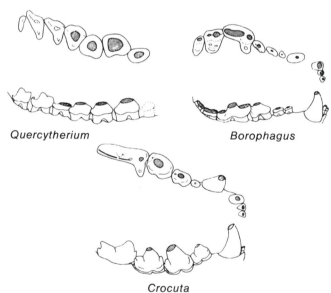

Quercytherium

Borophagus

Crocuta

Some carnivores have cheek teeth adapted to crushing bones or molluscs; these premolars have thick enamel and broad cusps that wear down to provide a crushing platform.

Carnivorous marsupials

Carnivores have evolved four or five times among marsupial stocks, three times in South America and twice in Australia; both continents were isolated during most of the Tertiary and placental carnivores only arrived during Plio-Pleistocene times. In the fossil record marsupials are distinguished from placentals on a number of osteological and dental features. For example, in marsupials the dentition is rarely and only partially replaced, they have only three premolars but four molars, and in carnivorous forms carnassial specialization frequently develops on three pairs of molars simultaneously.

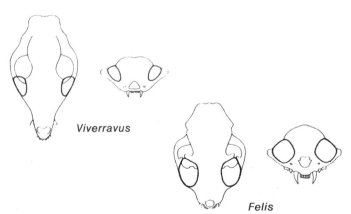

Viverravus

Felis

In the extinct and primitive carnivore *Viverravus* the orbits are small and set on the side of the face. In advanced carnivores such as the cat *Felis*, vision is very important, thus the orbits are large and are set on the front of the face to give stereoscopic vision.

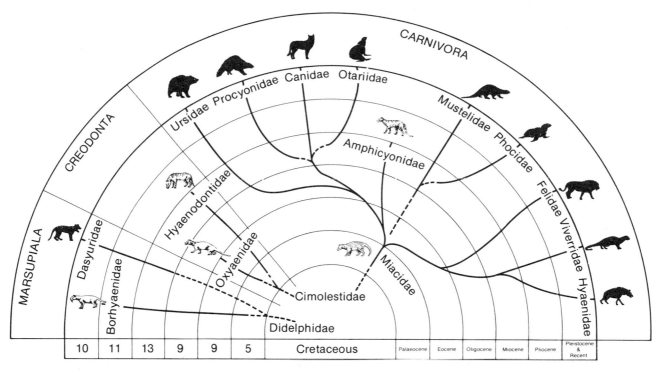

Relationships of the families of carnivorous mammals.

South American species

Didelphids are known from the late Cretaceous of South America and are still there as opossums, occupying small terrestrial omnivore–carnivore niches, though possibly never very abundant. The majority of carnivorous marsupials throughout the Tertiary in South America were borhyaenids; these are known from some 25 genera ranging from the Palaeocene through to Pliocene times. Most are from sites in Argentina, but some come from the far north in Columbia. Ecologically three major types can be recognized. There were small

to medium-sized omnivore–carnivore forms – the hathlyacynines; among these *Cladosictis* is well known from skeletal remains which show it to have been of similar size and limb proportions to an otter. The niches occupied by these hathlyacynines were taken on their extinction during the Pliocene by didelphids, mustelid and canid placentals.

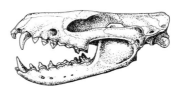

Skull of *Cladosictis* (13 cm long) with well-developed canines and carnassial teeth. (After Sinclair).

Restoration of *Cladosictis*, an otter-like borhyaenid marsupial from the Miocene of Patagonia.

Restoration of *Prothylacynus*, a wolverine-like borhyaenid marsupial from the Miocene of Patagonia.

The prothylacynines were medium to large-sized terrestrial omnivore–carnivores, resembling wombats and wolverines of today. Such was *Prothylacynus* from the richly fossiliferous volcanic ash Miocene deposits of Santa Cruz in southern Argentina. With the establishment of links with North America in the early Pliocene, raccoons and bears invaded South America and took over the niches occupied by the prothylacynines. The third ecological group was that of the large to very large terrestrial carnivore, filled by the borhyaenines, with, for example, *Borhyaena*. These were very varied with up to about a dozen genera which ranged in size from that of a small fox to some that were as large as a lion; one poorly known species appears to have had a skull that was about 60 cm long, which puts it in the size category of the largest hyaenodont

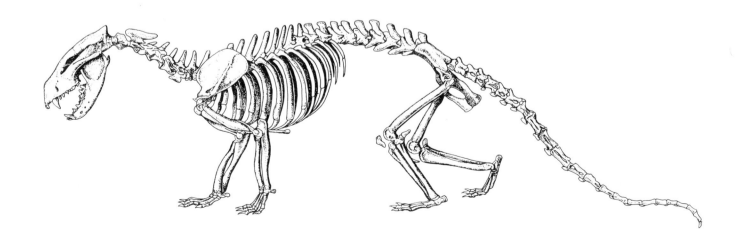

Restoration of the marsupial *Borhyaena,* from the Miocene of Patagonia; some of these were as large as a brown bear; the skull and dentition were hyaena-like.

(*Megistotherium*). Throughout much of the early Tertiary, the borhyaenines were the dominant predators on the large and varied ungulate faunas that evolved on the isolated continent. They were not, however, highly cursorial animals; they remained largely plantigrade, short legged, heavily built animals. However, their skulls were armed with large strong canines, with heavy crushing premolars and sharp, elongate carnassial

molars. During the Miocene, and increasingly more so during the Pliocene, the borhyaenines had competition as herbivore predators from the phororhacoid birds. These were large terrestrial birds that lived like latter-day dinosaurs and took over the large predator niche from the borhyaenids; they held on to it until the arrival of the placental dogs and cats in the Pleistocene.

Skeletons of *Prothylacynus* (head and body 80 cm) (opposite), and *Borhyaena* (head and body 1 m). Both these carnivorous marsupials had long tails, shortish limbs and were plantigrade or flat footed. (After Sinclair).

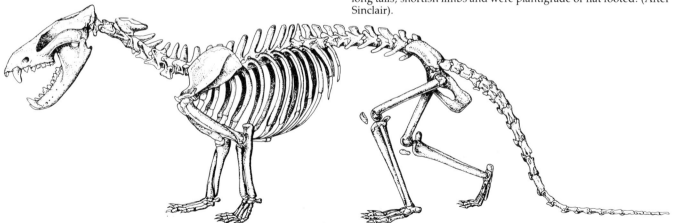

The third stock of South American marsupial carnivores included the spectacular sabretooth *Thylacosmilus* and a few other close relatives, known only from the late Miocene and Pliocene. However, *Thylacosmilus* even though its ancestry is unknown, was a very advanced sabretooth with a canine which was larger and longer than any developed in true cats. *Thylacosmilus* differed from true sabretooth cats in having a sabre that apparently was continually growing, rather like a rodent incisor. Nevertheless, this superb beast, along with the remaining borhyaenids, became extinct with the arrival in South America during the early Pleistocene of the true sabretooth cats.

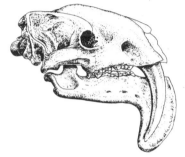

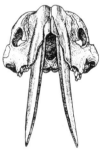

Skull of *Thylacosmilus* (20 cm long), the sabretooth marsupial carnivore. The upper canines are proportionately larger than in any placental sabretooth cat. There is a deep flange on the mandible to protect the sabretooth when not in use. (After Riggs).

Restoration of *Thylacosmilus* from the Pliocene of Argentina. In appearance it closely resembled the Plio-Pleistocene sabretooth cats of North America and the Old World.

Australian species

In the isolation of Australia, marsupials were the only land mammals, other than a few monotremes, until the arrival in Plio-Pleistocene times of rodents. While most of the diverse radiation was of herbivorous forms, these were kept in check by the predations of dasyurids, a family that has parallels with both the didelphids and the borhyaenids. Dasyurids range in size from that of a shrew to that of a wolf. They are known as far back as early Miocene and many are still extant, of which the cat-like Tasmanian devil, *Sarcophilus* is perhaps the best known. Remains of dasyurids are not uncommon in Pleistocene bone caves in many parts of the continent, and cave floor debris sometimes includes chewed bone fragments, the remnants of their meals.

The largest Australian marsupial carnivore was a strange beast named *Thylacoleo*, whose closest relatives are the living phalangers, a family of fruit, flower and plant-eating arboreal mammals. Marsupial lions are now known from Miocene to Pleistocene in Australia; they reached the size of a leopard and were probably quite lion-like in appearance. The head was short; powerful jaws were armed with strong canine-like incisors and exceptionally long (5 cm) carnassial blades extending across two teeth ($P_3 + M_1$). For many years it was doubted that this animal was a carnivore; because apart from its dentition, it is in other ways very similar to phalangers; perhaps it was adapted to some specialized vegetarian diet. Recently, however, it has become possible using high power microscopes to differentiate between scratches left on teeth by masticating vegetable food and scratches made by chewing meat. From this study it is clear that *Thylacoleo* was indeed a meat-eater, preying on wombats and diprotodons and other large slow moving marsupial herbivores.

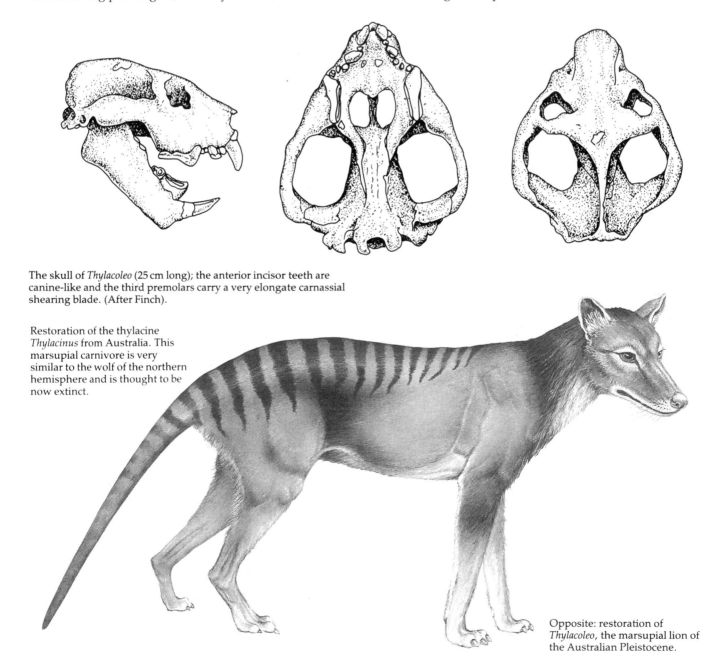

The skull of *Thylacoleo* (25 cm long); the anterior incisor teeth are canine-like and the third premolars carry a very elongate carnassial shearing blade. (After Finch).

Restoration of the thylacine *Thylacinus* from Australia. This marsupial carnivore is very similar to the wolf of the northern hemisphere and is thought to be now extinct.

Opposite: restoration of *Thylacoleo*, the marsupial lion of the Australian Pleistocene.

Creodonts

The dominant carnivores around the world throughout the early Tertiary, except in Australia and South America, were the members of the extinct order Creodonta, of which some 50 genera are known in two families – the Oxyaenidae and the Hyaenodontidae. Creodonts are distinguished from true carnivores (Carnivora Vera) on a number of osteological characters (shorter limbs, unfused wrist bones, cleft claw bone and lack of an ossified auditory bulla) and dental characters, of which the most important relates to the carnassial teeth. In living carnivores, the carnassial pair is always P^4/M_1; in creodonts it varies, but usually involves two pairs of teeth, most commonly M^{1+2}/M_{2+3}. In primitive forms the blade is angled diagonally across the jaw, but in later advanced forms it is longitudinal as in modern carnivores.

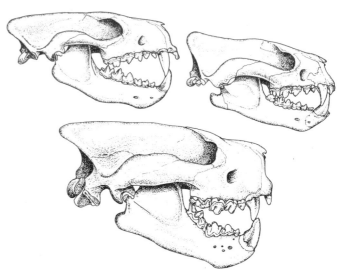

Skull of *Sinopa* (15 cm long) from the Eocene of North America; a fox-like creodont. (After Matthew).

Skulls of three creodonts from the Eocene of North America. All have short faces, well-developed canine teeth, carnassial molars and strong closing jaw muscles. *Patriofelis* skull (bottom) 30 cm long; others to same scale. (After Denison).

Restoration of three creodonts. *Sinopa* (left) was fox-sixed, *Patriofelis* (top) was bear-sized and *Oxyaena* was about the size of a glutton.

Oxyaenids

The oxyaenids are confined to the Palaeocene and Eocene of North America and Eurasia, with about a dozen genera, which basically find parallels among the mustelids and felids. Most were medium-sized animals with short robust limbs, plantigrade feet and a very long tail. A few grew very large and *Sarkastodon* was as large as the largest bear. One form, *Palaeonictis*, was very cat-like and another *Patriofelis* was rather bear-like.

Hyaenodontids

The hyaenodontids ranged over the northern continents and reached Africa. They also survived until the late Miocene in Africa and Asia. Three distinct stocks can be recognized within the hyaenodontids; there were sabretooth forms – the machairoidines – though these are limited to the late Eocene of North America and do not seem to have been very abundant. The second stock were the proviverrines, which included civet and dog-like forms. They also include *Quercytherium*, a medium-sized genus from France with heavy crushing premolars reminiscent of the living sea otter; perhaps *Quercytherium* also ate crushed molluscs.

Restoration of *Hyaenodon*. The species of *Hyaenodon* varied from stoat to hyaena in size. They were very successful predators and probably also carrion feeders. Deinotheres in background (see Chapter 10).

The best known of the three stocks was the hyaenodontine subfamily. These were mostly large terrestrial carnivores, occupying the niches which today would be filled by canids and hyaenas. The genus *Hyaenodon* is known from many species in North America, Eurasia and Africa, ranging from some the size of a stoat to others as large as a hyaena. One form, *Megistotherium* had a skull which was 65 cm long and must have been the largest carnivore that has ever lived. The hyaenodontines had elongate limbs, digitigrade feet and were very agile. The skull was very large in proportion to the body and some appear to have continued growing throughout life. The large strong canines, the heavy premolars and the efficient slicing carnassial teeth leave no doubts about the massive power of these formidable beasts; hence the question why did they become extinct.

Creodonts and living carnivores can both be traced back into the earliest Tertiary, but it was the creodonts which dominated the early Tertiary carnivore faunas. When we compare their anatomy with that of the contemporary dogs and cats, they seem equally matched. In both stocks the transformations of the limbs to faster running took place about the same time. In both stocks there were forms with stereoscopic vision, and brain development does not seem to be significantly different. The extinction of the creodonts and their replacement by our living carnivores is an unexplained mystery.

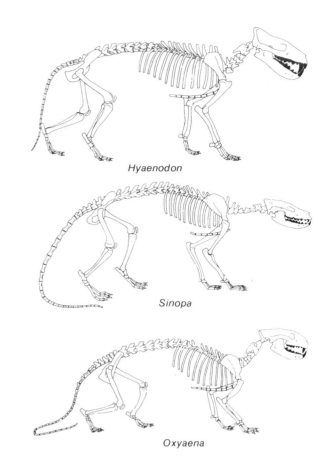

Skeletons of three creodonts: *Oxyaena* (after Osborn), *Sinopa* (after Matthew) and *Hyaenodon* (after Scott & Jepsen). All have long tails, short limbs and well-developed canine teeth.

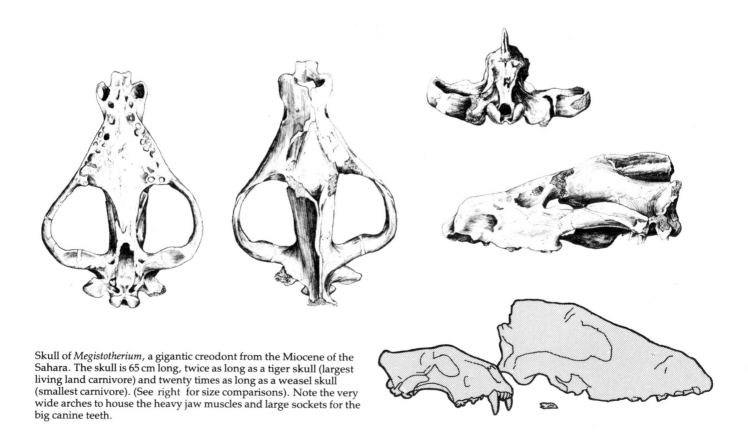

Skull of *Megistotherium*, a gigantic creodont from the Miocene of the Sahara. The skull is 65 cm long, twice as long as a tiger skull (largest living land carnivore) and twenty times as long as a weasel skull (smallest carnivore). (See right for size comparisons). Note the very wide arches to house the heavy jaw muscles and large sockets for the big canine teeth.

Carnivora Vera

Some confusion inevitably arises when the word carnivore is used in two senses – for a taxonomic order of mammals and for mammals which feed on meat. The two are not synonymous though there is a great deal of overlap. Most members of the order Carnivora do eat meat (but not badgers, aardwolf or pandas). Most meat-eating mammals are members of the order Carnivora, but not the carnivorous marsupials or the extinct creodonts. Hence the use of the term Carnivora Vera which means 'true carnivore', to distinguish the taxonomic order from other mammals adopting carnivory. The fossil record makes it clear that both the Creodonta and the Carnivora can be traced back separately into the earliest Tertiary, and both probably share a common origin in early Palaeocene or latest Cretaceous from an insectivorous stock not unlike *Cimolestes*.

Characters which distinguish marsupials and creodonts from the Carnivora Vera are noted above.

Within the true carnivores three major streams or superfamilies can be recognized, which we may call the musteloid, the canoid and the viverroid. The basis for this grouping is the structure of the bony auditory bulla in the middle ear, the elements which go to making up the bulla and the presence or absence of a bony septum dividing the bulla. Of the families that make up the order, only one – the Amphicyonidae – is extinct.

The Carnivora Vera remained relatively minor components in the faunas throughout the Palaeocene and the Eocene, the creodonts providing the main carnivore stocks. Then in later Eocene times there arose ancestral canid (dog) and viverrid (civet) lineages, and possibly also the mustelid (stoat) lineage. The early Tertiary carnivores which preceded the modern lineages are often grouped together into the family Miacidae, but as this is not a monophyletic group it is best not used. The genus *Miacis* is well known from the Eocene deposit at Messel in Germany; it was a small arboreal mammal very like a pine marten. It is preserved in a coal deposit, which indicates the former presence of tropical forest vegetation. The family Mustelidae (stoats, martens, weasels, badgers, skunks and otters) could have evolved from such an animal, but the fossil record of the group is poor, save for the otter branch of the family and from these evolved the seals.

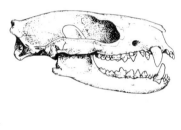

Skull of *Vulpavus* (10 cm long); this miacid skull resembles that of the living marten. (After Matthew).

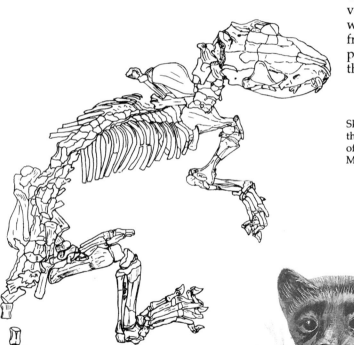

Skeleton of *Miacis* with very long tail and limbs that could grasp branches. Head and body 15 cm. (After Springhorn).

Restoration of *Miacis*, a miacid from the mid Eocene of Messel in Germany. *Miacis* was an arboreal pine marten-like animal living in tropical forests.

Canoids

CANIDS, THE DOGS   The dogs belong to the family Canidae which forms the core of the superfamily Canoidea. They first appear in late Eocene times and are at that time more like civets than dogs in external appearance. Such was *Pseudocynodictis* from the Oligocene of North America, where the dogs had their centre of evolution, with occasional representatives in Europe and Asia; not until the Plio-Pleistocene did they spread to other continents. Canids, comprising foxes, jackals, wolves and dogs are very successful general purpose carnivores, able to adapt to a wide variety of habitats and a great range of foods. They have a good carnassial dentition and retain substantial crushing teeth behind these, so that they can subsist on a vegetarian or omnivorous diet. They have good stereoscopic vision, keenly developed sense of smell, wide frequency range of hearing, long limbs and digitigrade feet that enable them to run fast. This combined with their stamina enables them to range over very extensive territories in search of food. On top of all this they are highly intelligent and are mostly social animals, taking advantage of their numbers in hunting and rearing the young. In North America during the Miocene the canids evolved a side branch, the borophagines or hyaena-dogs; these large canids had very powerful jaws and large heavily enamelled premolar teeth that enabled them to crush bone as the hyaenas do; they filled a niche left vacant by the true hyaenas which never made a successful impact in the New World.

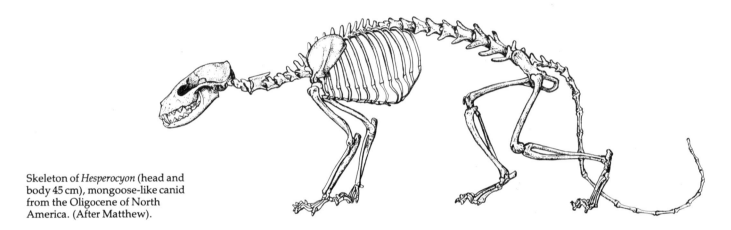

Skeleton of *Hesperocyon* (head and body 45 cm), mongoose-like canid from the Oligocene of North America. (After Matthew).

Restoration of *Hesperocyon* from the Oligocene of North America; this animal was probably very similar in appearance to the meerkat, a living mongoose, but from its dentition and ear structure, it can be shown to be related to dogs.

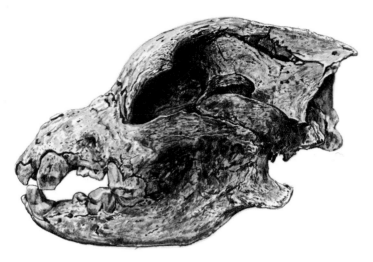

Skull of *Borophagus* (18 cm long), a large hyaena-like bone crushing dog of the North American Pliocene. (After Matthew & Stirton).

AMPHICYONIDS, THE BEAR-DOGS The amphicyonids or bear-dogs were an abundant, varied and widely successful stock. Their origins go back into the late Eocene, they diversified considerably in the Miocene as the creodonts declined, and they survived into the Pliocene in Eurasia and Africa when their place was taken by canids. In body shape they resembled bears – large, short limbed, flat footed; the face and dentition resembled wolves, though many were much larger than wolves. *Amphicyon giganteus* from European Miocene was as large as a tiger. They would have been redoubtable hunters of the abundant game that roamed warm temperate belts of the world during mid Tertiary times.

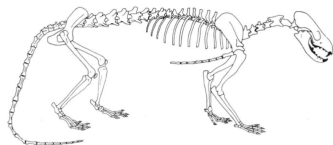

Skeleton of *Daphoenus* (head and body 1 m); note the long tail, long limbs and digitigrade feet. (After Scott & Jepsen).

Skull of *Daphoenus* (20 cm long) is rather badger-like in appearance with broad crushing molar teeth, weakly developed carnassials and strong bony crest over the brain case. (After Scott & Jepsen).

Restoration of *Daphoenus*, a primitive amphicyonid or bear-dog from the Oligocene of North America. This genus was lightly built like a greyhound dog.

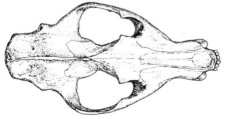
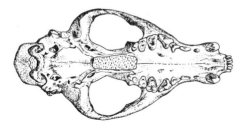

Restoration of *Amphicyon*, a large tiger-sized carnivore with bear-like skeleton and dog-like dentition; it was common in Europe during the Miocene.

Restoration of *Agriotherium*, a primitive bear that lived in Eurasia in the Miocene and survived in South Africa until the Pliocene.

URSIDS, THE BEARS   The true bears or ursids differ little from the amphicyonids in their skeletons, but dentitions clearly separate them. Bears have totally lost the carnassial function of the P^4/M_1 and developed instead an extensive crushing area on the posterior teeth. Their diet is very varied and includes vegetarian components; some have become very large and they include the largest living land carnivore, the kodiak bear with a weight of around 800 kg. There is also an aquatic species, the polar bear and in the Pleistocene there was a large cave bear. Bears have been very successful on the northern continents, preferring temperate to tropical climates. During the Pleistocene they invaded South America where they still persist. However, it is Africa which has largely but not totally defeated them. We have a record of brown bears in the Atlas mountains during the Pleistocene, and even surviving into modern times; it is easy to conceive how they could have crossed from Spain at the Straits of Gibraltar during a low sea-level period of the Pleistocene. The second record is not so easily explained; skulls and other remains of the bear *Agriotherium* have been found at Langebaanweg in South West Africa; the site is dated earliest Pliocene, about 5 million years ago. There are no other records on the whole of the African continent of bears, and how and when it got to South West Africa is a mystery.

Restoration of *Hemicyon* a very successful ursid from the Miocene of the northern hemisphere. *Hemicyon* (literally half dog) was probably more active and carnivorous than true bears, but more sturdily built than most dogs.

AILURIDS, THE PANDAS   Anatomically and physiologically very close to the ursids are the Ailuridae, the pandas. Indeed there is considerable uncertainty whether the giant panda is or is not a bear. Fairly closely related to the pandas but separated from them at family level are the raccoons; this fairly small group is limited to the Americas where it fills an arboreal omnivore niche. They made a particularly successful invasion of South America in the early Pliocene; one form *Chapalmalania* closely paralleled the giant panda in size and appearance. The final branch of the canid lineage was that of the otariids and odobenids, the sealions and walruses dealt with under aquatic mammals.

Restoration of a cave bear (*Ursus spelaeus*). This very large species was abundant in Europe during the Pleistocene ice ages and many died while hibernating in caves.

Restoration of *Chapalmalania*, a giant panda-like raccoon from the Pliocene of South America.

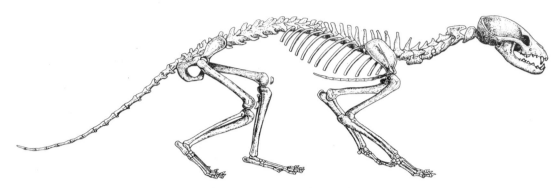

Viverroids

The third major superfamily of carnivores, the viverroids are characterized by having a septum in the auditory bulla. Within the superfamily are three families – viverrids (civets and mongooses), hyaenids (hyaenas) and felids (cats).

VIVERRIDS, THE CIVETS AND MONGOOSES The viverrids are the oldest stock, with an ancestry going back into the Palaeocene. They have diversified greatly to become the most prolific carnivore family; today there are no less than 36 genera; their distribution is mainly in the tropics of the Old World – they never invaded the New World. They are all medium-sized mammals with an average of around 4 kg and in the tropics are the most important rodent predators, by virtue of both their diversity and their numerical abundance. In spite of this they have a poor fossil record. *Kanuites* is a genus from the Miocene of Africa known from complete skeletons; in life it resembled a genet. Like the genet it was an opportunistic feeder – fruit, berries, grubs, insectivores, frogs and anything that chanced its way. From the viverrids evolved two other stocks, the hyaenas and the cats.

HYAENIDS, THE HYAENAS The hyaenids began life in the early Miocene as civet-like carnivores and such an animal was *Ictitherium* from the late Miocene of Greece. They were distinguished from viverrids mainly by the absence of a visible septal division of the auditory bulla; hyaenas still have a septum, but it no longer divides the bulla in two equal halves, one part having become very much larger than the other. The second stream of hyaenids produced what we typically think of as hyaenas; the large, fearsome bone-crushing forms. As we have seen they were not the only stock to live in this way, but they are the only survivors today. The jaws of a spotted hyaena (*Crocuta*) are so powerful that it can exert a pressure of some 800 kg on its premolar teeth (equivalent to 3 tonnes cm^{-3}).

Skeleton of *Kanuites* (head and body 50 cm), a primitive mongoose from the Miocene of Kenya.

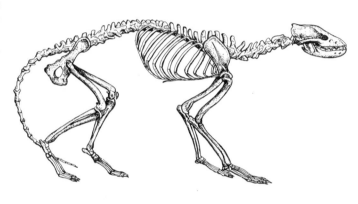

Skeleton of *Ictitherium* (head and body 85 cm), a civet-like hyaena from the late Miocene of Greece. (After Gaudry).

FELIDS, THE CATS The felids are the ultimate in carnivore evolution – the kings of the beasts; large, impressive, fast, ferocious, awesome and deadly. And at the more humble level – the domestic cat – independent and yet still a beloved domestic favourite. Indeed the pussy cat is much more typical of the family of cats than the lion. The majority of felids are small to medium-sized and live like the cat on a diet of rats and mice, as they have done for millions of years. The keen senses, quick intelligence, nimble feet, fast movements and sharp teeth all combine to make cats extremely efficient predators. They range worldwide except for Australia (where they have been introduced by man).

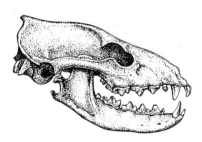

Skull of *Ictitherium*.

Opposite: restoration of *Pachycrocuta brevirostris*, a gigantic hyaena that was widespread across Eurasia during the mid Pleistocene; it stood 1 metre at the shoulder – as large as a lion.

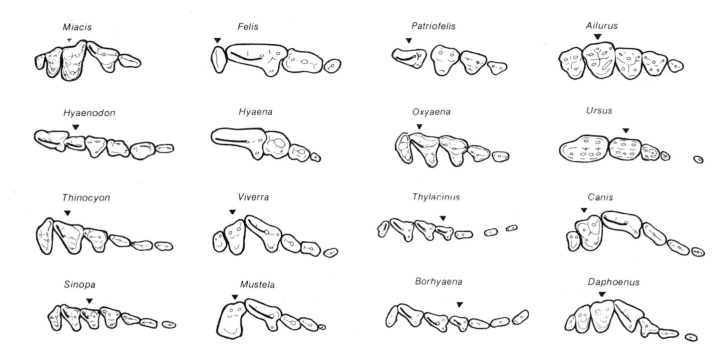

Miacis · Felis · Patriofelis · Ailurus

Hyaenodon · Hyaena · Oxyaena · Ursus

Thinocyon · Viverra · Thylacinus · Canis

Sinopa · Mustela · Borhyaena · Daphoenus

Occlusal views of the upper cheek teeth of representative carnivores; the carnassial blade is indicated with a heavy line; a black triangle is placed over M^1.

In the fossil record of the felids the limelight falls inevitably on the sabretooth cats, sometimes misnamed sabretooth tigers (for they were not closely related to tigers). As we have seen the evolution of a sabretooth occurred several times; among marsupials and creodonts and at least twice within the felids.

Enlargement of the upper canines to form sabretooth canines has led to numerous other modifications in the skull and jaws. The face is always short, the neck musculature strong, especially that for pulling the head downward. The lower jaw is modified to open to a wide gape, while the musculature still retains the ability to

Relationships of subfamilies, tribes and genera within the cat family Felidae. (After Thenius).

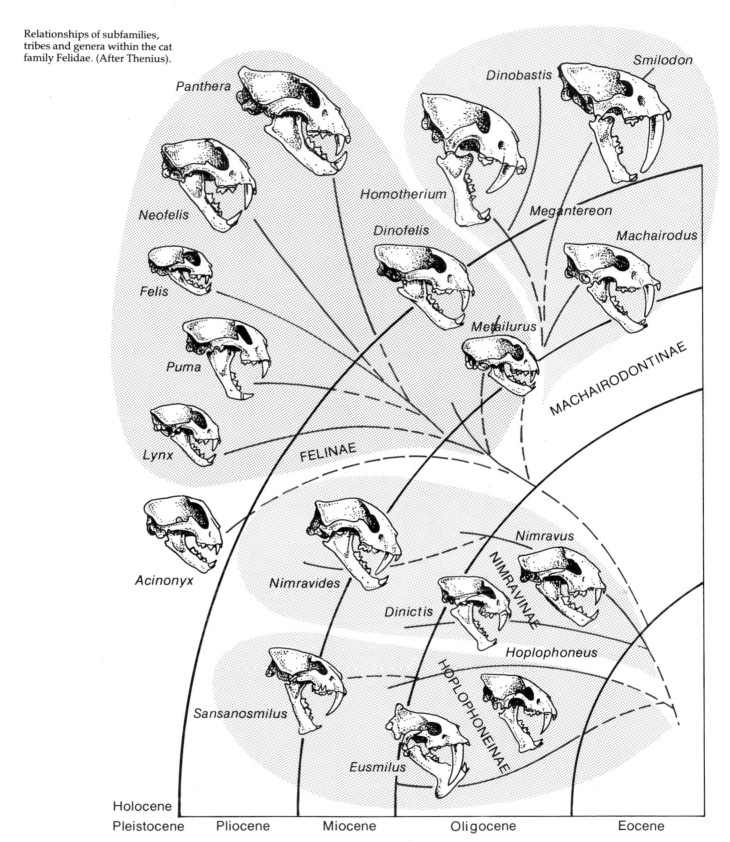

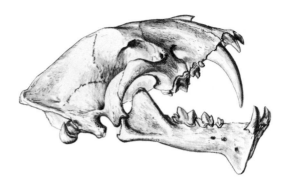

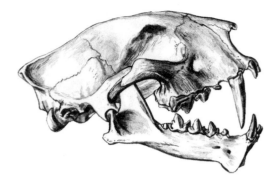

Skull of *Haplophoneus* (16 cm long), representative of a lineage of false sabre-tooth cats which flourished during the Oligocene in North America. (After Scott & Jepsen).

Skull of *Dinictis* (14 cm long); the cheek teeth are much reduced, only the M^1 and M_{1+2} are present. The upper canine is thin, elongate and curved in the form of a short sabre. (After Scott & Jepsen).

exert a powerful bite with the carnassial teeth. The upper canines besides becoming more elongate, also develop a distinct backward curve, become laterally compressed (a compromise between circular cross-section for strength and a wafer thin cross-section to minimize resistance). The anterior and more particularly the posterior border acquire a serrated edge (as in a bread knife to give a good tearing edge for unlike a knife a tooth cannot be sharpened). Sabreteeth are efficiently adapted for stabbing actions, into skin and flesh but not bone; for this reason it is likely that they would have been used to penetrate the throat and lower neck rather than used on the upper neck where they might be broken against cervical vertebrae. A secondary function would have been to slice and tear up muscle, but this would only be possible after the initial stab which killed the victim. Unhappily no sabretooth survives today; the last became extinct in the late Pleistocene but must have been well known to our Palaeolithic ancestors.

Restoration of *Dinictis* from the Oligocene of South Dakota. This species was about the size of a serval cat.

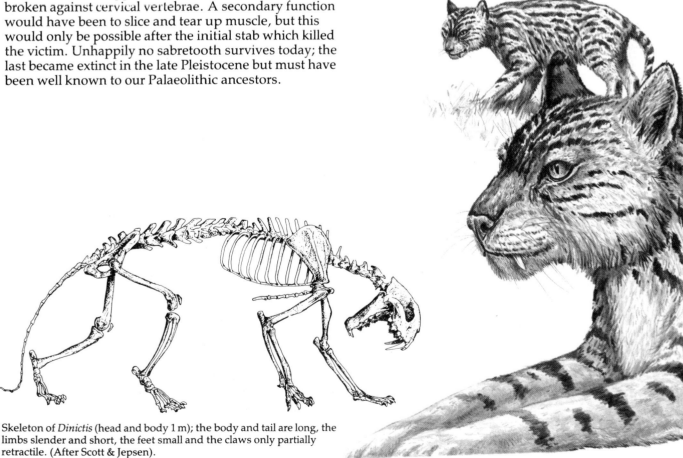

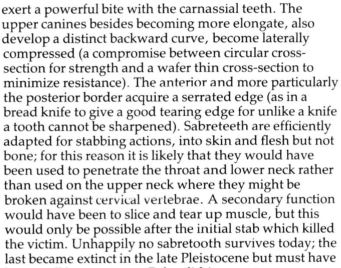

Skeleton of *Dinictis* (head and body 1 m); the body and tail are long, the limbs slender and short, the feet small and the claws only partially retractile. (After Scott & Jepsen).

Restoration of the cave lion
(*Panthera spelaea*). This
Pleistocene lion is so similar to the
living lion that it may be the same
species. It is found abundantly in
caves all over Europe, surviving
into historic times in the Balkans.

Restoration of *Smilodon*; this
sabretooth cat was larger than the
dirktooth cat, reaching the size of
a lion. It had very elongate sabres
and its remains are abundant in
the Rancho La Brea tar pits of
California where in the
Pleistocene it preyed on bison
and mammoth trapped in the tar.

Restoration of *Megantereon*, the dirktooth cat. This successful cat had very elongate upper canine teeth and was widespread in the Pliocene ranging across Eurasia, North America and as far south as South Africa.

Restoration of *Homotherium*; this large scimitar tooth cat was abundant during the early and mid Pleistocene in Europe. Its canine teeth are shorter and more flattened than in the sabretooth cat. The fore limbs are greatly elongated and the hind limbs are short and plantigrade.

CHAPTER 7

Paddlers and swimmers

Moving through water

Amongst the greatest achievements of mankind must surely rank the invention of the sail. With this simple innovation man has travelled the oceans of the world; for a truly terrestrial mammal and one of the very few innately unable to swim, this was one of the most remarkable accomplishments of his early civilizations. How strange that nature never invented the sail, unless the jellyfish *Velella* qualifies. Man's recent invention of the screw propeller is, like the wheel before it, not an option available to other vertebrates. Basically all aquatic mammals row or paddle, using their paired limbs to row and their tail to paddle. These have become very sophisticated in fully aquatic mammals like sealions and whales, with hydrofoil section and great economy of energy in utilizing both primary and recovery strokes for propulsion.

Rowing

In rowing there are one or more pairs of oars, held with the shaft almost horizontal as the broad blunt blade dips into the water; this shape creates the maximum

Restoration of *Allodesmus*, a relative of sealions from the Miocene of the Californian coast. These early sealions probably used their large eyes to track down and capture fish in shallow waters. Their ears would have functioned well on shore but were not adapted to the pressure of deep diving.

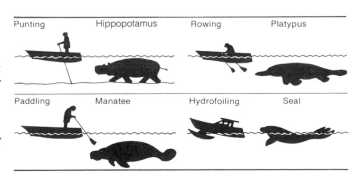

The four ways animals can propel themselves through water: a hippopotamus punts or walks on the bottom; a seacow paddles by flipping its tail up and down; a platypus rows with sweeps of its broad webbed feet; a seal hydrofoils using its flippers.

turbulence and resistance. Technically rowing refers to one man per oar, while in sculling one man has a scull or oar in each hand. Paddling is similar to rowing except that there is often only one oar, held vertically, and used either from the side or rear of the boat. (In punting a long pole reaches to the river bed and the action is essentially similar to walking.)

We can see an evolutionary sequence from the simple small rowing boats so popular at seaside resorts to the highly efficient rowing machines competing in the Henley Royal Regatta. Small boats have short oars and the rowlocks attached to the sides of the boat, which is usually rather broad, partly for stability and partly to give a reasonable distance between the fulcrum (rowlocks) and the rower's grip, the distance being the moment arm. In the racing boats the oars have 4-metre-long shafts and the rowlocks are outrigged well clear of the narrow streamlined boat; all this gives much greater mechanical efficiency to the stroke and resistance offered by the boat's contact with the water is minimized. Mammals that row, such as otters with short legs, are closer to the seaside rowing boat in design. They do not have the option of extending the fulcrum (shoulder and hip joint) beyond the body contour. To substantially improve propulsion efficiency animals employ a different tactic, they use the hydrofoil.

The hydrofoil

A sealion or seal flipper illustrates a hydrofoil. The flipper is thicker along the leading edge, thinning backward, tapering toward the distal end and has a smooth surface. The contour can be altered through muscular action. Such a hydrofoil enables the animal to take advantage of the properties of laminar flow of water over the flipper. In swimming the flipper moves through a figure of eight; feathering – slight changes of angle of attack – enables it to achieve lift and propulsion on both the up and down strokes.

Manatees have a very large broad horizontal fan-

A monk seal using its front flippers as hydrofoils to fly through the water.

shaped tail which is used as a paddle. Whales also use their tails but these are more fish-like in shape; the tapering tail is terminated in a pair of horizontal flukes which behave as hydrofoils. Thus as the tail flukes are alternately flexed upward and downward, thrust is provided on both strokes. Fish achieve the same goal by sideways movements of their tail fins.

Living in water

The enormous challenge of adapting to aquatic life has been tackled in varying degrees by many mammals. Otters are expert swimmers, highly mobile in water while retaining considerable agility on land. Sealions are more aquatically adapted yet are still able to move about on land. Whales are totally aquatic and die rapidly if not in their watery environment.

The fundamental problems are to move efficiently through water whilst locating and ingesting food, to stay in water away from predators, to keep warm, to breathe and to produce young. The more adapted a mammal is to aquatic life the more fully all these functions are performed in the water. Virtually everything in the anatomy and physiology of whales is altered to accommodate them to the environment of fishes, from which they are indirectly descended. The fossil record enables us to trace the evolution of these adaptations in the skeleton; from the bones and teeth we can infer a good deal about the brain and diet, and even learn something about their respiration and breathing.

The total number of aquatic mammal species is large; the actual figure depends on whether we include the 'paddlers', that is species that get their feet wet in foraging for food. Mammals often swim across rivers and even lakes without any specialized adaptations. A list of living aquatic mammals would certainly include the following:

Monotreme	Platypus
Marsupial	Water opossum
Insectivores	Water shrews, desman
Rodents	Beaver, capybara, coypu, water rat, water vole
Perissodactyls	Tapirs
Artiodactyls	Water hog, hippopotamus, water deer, moose, water buck, water buffalo
Sirenians	Dugong, manatee
Carnivores	Otters, polar bear, water mongoose, seals, sealions, walrus
Cetaceans	Whales, dolphins, porpoises

Aquatic invasions by mammals could never have been an easy option to competition on land; as well as all the problems of coping with the strange conditions of a new environment, they had to compete with the fishes, reptiles and birds that were there before them – sharks, turtles, sea snakes, sea lizards, crocodiles, swimming and wading birds. Despite the obvious advantages of an aquatic environment for preservation, the fossil record of aquatic mammals is no better than that of land mammals and in the critical early stages of the evolutionary adaptations it is worse.

Herbivores

Aquatic mammals can conveniently be grouped into vegetarians and flesh eaters. The vegetarians tend to be less specialized, browsing freshwater plants or offshore seagrasses; none cruise the oceans and no speed is required to catch food; most are large and so are in less danger from predators. There are few aquatic herbivores in comparison with the very numerous carnivores, otters, seals, whales, etc, which is surprising since on land the situation is reversed. Water hog, water deer, water buck and water buffalo are marginally aquatic, sometimes feeding on aquatic vegetation. Tapirs more often feed on aquatic plants. Of aquatic herbivores this leaves only hippopotamuses and seacows, to which we must add the extinct seahorses, the desmostylids.

Hippopotamuses

There are two living hippos – the Pygmy, a forest dweller in West Africa, and the large *Hippopotamus amphibius*, which is still found in many parts of Central Africa, but formerly was widespread on the continent, with relatives in the Plio-Pleistocene of Eurasia. Hippos have a number of aquatic specializations – they are very large, up to 3 tonnes, and have short stocky legs and large feet. They can traverse land but prefer to stay within easy reach of water where their great bulk is more readily supported. They are slightly denser than water and so have no problem in staying down and walking on the bed of rivers and lakes. Hippos are not ruminants but have two accessory stomach sacs (like

peccaries); their digestion is very slow and very efficient – their daily food requirement is only about 1·3 per cent of their body weight, about half the amount cattle require. Hippos graze on grasses in rivers and lakes, including papyrus plants, and they are very partial to lotus, a water-lily abundant in the Nile. They have large incisor teeth (either two or three pairs depending on the species), enormous shearing canines and a battery of high crowned cheek teeth. The molar teeth are characterized by having four trefoil cusps.

EARLY HIPPOS   The fossil record of early hippos is poor; on the basis of tooth fragments from Kenya, they probably arose in mid Miocene times. Their most likely ancestors are either members of the extinct anthracotheres (large pig-like mammals) or Miocene peccaries. Most species are about the size of the pygmy hippo and like it were probably forest dwellers. On Crete and Malta there were dwarf Pleistocene species. A line of large aquatic hippos can be traced in the Olduvai Gorge succession of Tanzania. As we follow the beds up from Bed I (1·8 MYA) to Bed IV (0·3 MYA) the hippos get larger, the face longer and the eye sockets more elevated on stalks. The final gigantic *H. gorgops* had periscopic eyes, so that while the whole body remained submerged, it could enjoy a panoramic view of its surroundings.

Restoration of *H. gorgops* from the Pleistocene papyrus swamps of East Africa. The eyes were elevated on stalks and could be used as a periscope to view the world while the hippo remained submerged.

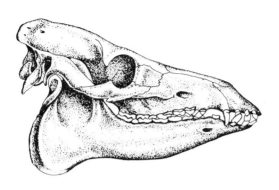

Skull of anthracothere *Bothriodon* (45 cm long) from the Oligocene of Europe and North America; these extinct pig-like animals may be related to hippo ancestors. (After Scott).

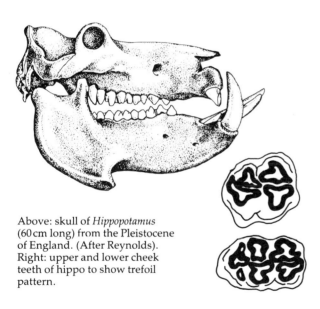

Above: skull of *Hippopotamus* (60 cm long) from the Pleistocene of England. (After Reynolds). Right: upper and lower cheek teeth of hippo to show trefoil pattern.

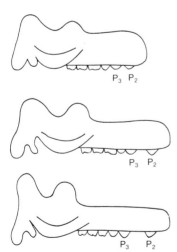

Sequence of changes in the shape of the hippo skull as followed through the Pleistocene in East Africa. The series ends in *H. gorgops*, a hippo with periscopic eyes. (After Coryndon).

Seacows

Manatees (three species of *Trichechus*) and dugongs (one species of *Dugong*) or seacows as they are collectively known, are the sole surviving representatives of the order Sirenia. Manatees live in tropical rivers opening into the Atlantic Ocean; they are also found around the coasts and islands of Central America. Dugongs inhabit tropical coastal seas around the Indian Ocean. Seacows are totally aquatic, though manatees have been known to rest on sand banks in a river.

Sirenians are huge shapeless animals with a large tail, small front flipper, no hind flipper, short neck, no external ears, small eyes and a strange blunt mouth with thick bristly overhanging lips. They have been recorded in excess of 4 metres long and weighing 900 kg. Their skeleton reveals only a vestige of the pelvic girdle. The fore limb is used for manoeuvering and for walking on the river bed. The large horizontal tail is the sole means of swimming. It is a very broad spatulate paddle in manatees and enormous relative to the size of the animal; looking at a manatee from above, the tail accounts for about one third of the surface area. In dugongs the tail is divided into two horizontal fins. Seacows swim slowly; 10 kh^{-1} has been recorded and they can stay submerged for five minutes and perhaps on occasions for as long as 15 minutes.

Many skeletal characters betray aquatic specializations. The ribs are enormously thick and dense, especially in the manatee; this adds weight and enables it to stay under water without effort. This phenomenon of bone thickening is known as pachyostosis and is very rare among vertebrates. Their metabolic rate is slow and they ingest about 5 per cent of their body weight daily. Manatees feed on aquatic plants, especially water hyacinth (*Eichornis cressipes*) and in this valuable activity manatees keep water channels open. Dugongs are bottom feeders on seagrasses and seaweeds; green algae are extensively grazed, but are usually secondary to the seagrasses. A close correlation in the distribution of seagrasses and dugongs can be traced back to Eocene times.

The anterior part of the snout of the dugong is down-turned. Food plucked with the overhanging lips is mashed between bristle pads attached like two scrubbing brushes just inside the mouth; the cheek teeth are small, cylindrical, single rooted and without enamel capping. In the front of the upper jaw two incisor tusks are imbedded; only the tips erupt through the bone. Manatees lack the down-turned snout and also the tusks; they have a long battery of bilophodont molar teeth on each side of each jaw, which are continuously replaced in escalator fashion; the teeth erupt at the back, migrate forward and drop out at the front. The nostrils of seacows are high on the snout, the small eyes lack a lacrimal gland and there is no external ear. The brain is small and only weakly patterned with a few folds.

A third genus of seacow, *Hydrodamalis*, was discovered by the German naturalist Georg Stellar in 1741 off Bering Island in the north Pacific. This vast mammal reached around 8 metres in length and weighed about 6 tonnes. The mouth was completely toothless and Stellar reported that it fed only on seaweeds. Sadly it was hunted to extinction just 27 years after its discovery. Stellar's seacow was unique among mammals in feeding exclusively on seaweed; with such a food abundantly available in the seas of the world throughout the whole of Phanerozoic time, it is indeed strange that so few vertebrates have ever exploited the resource. On the Orkney islands in Scotland there is a population of sheep that have adapted to a diet of seaweed, which demonstrates that it is possible for a terrestrial herbivore to adapt easily and quickly to such a diet.

ORIGIN AND EVOLUTION   The first clear evidence of sirenians dates from the mid Eocene and their evolutionary success at this time is tied in with two other events, the evolution of the Tethys seaway and the evolution of seagrasses. A broad seaway stretched east–west at around 20 – 30° north from Spain to China, with a similar sea around the Caribbean region. Deposits from this belt have yielded a rich fauna of foraminifera (protozoa with shells), some of which are known to have preferences for dwelling among seagrasses. The seagrasses are marine flowering plants (angiosperms), related to terrestrial grasses and the only seed-bearing plants successfully to invade the seas. They root into sediment, usually in waters less than 12 m deep, in both temperate and tropical seas. Seagrasses can form extensive sea-meadows and for dugongs are their major food source. The Eocene sirenians include both dugongids and trichechids (manatees), and they were already highly specialized to aquatic life. The dugongid *Eotheroides* from the Tethys had thickened bones, no hind limbs, bilophodont teeth and a down-turned snout with incisor tusk. The trichechid *Prorastomus* from the Caribbean had a less specialized skull and we do not have any postcranial skeleton.

The origin of the sirenians (like that of whales) is still quite unknown. On dental characteristics – bilophodont molars and the evolution of tusk-like incisors – a common ancestry of sirenians and proboscideans (elephants) in the Tethyan area is a distinct possibility. During the early Miocene, the European Tethys was rich in sirenians; near complete skeletons of *Metaxytherium* are known. The Caribbean and Central America continued to have dugongids throughout the Tertiary. The most striking forms were those that broke away from the tropical zones and penetrated the cold waters along the north Pacific coasts of America, becoming surface feeders on kelps. Domning has chronicled the gradual transitions within the hydrodomaline lineage from the early mid Miocene *Dusisiren* to the recently exctinct *Hydrodomalis gigas*: this involved major size increase (doubling length from 4 to 8 metres), increased buoyancy (with reduction of pachyostosis), loss of teeth and phalanges.

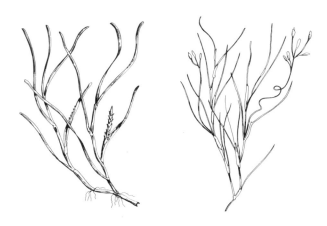

Seagrasses; these are monocotyledonous angiosperms, that is, true flowering plants related to grasses and they are unique in having adapted to living in the sea like seaweeds. Left *Zostera* and right *Ruppia*.

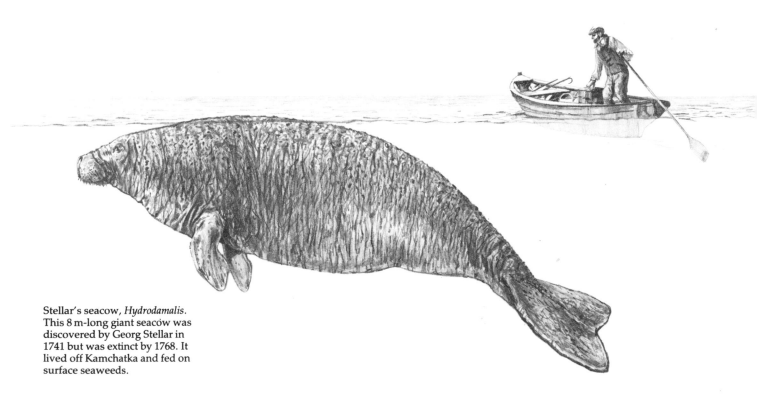

Stellar's seacow, *Hydrodamalis*. This 8 m-long giant seacow was discovered by Georg Stellar in 1741 but was extinct by 1768. It lived off Kamchatka and fed on surface seaweeds.

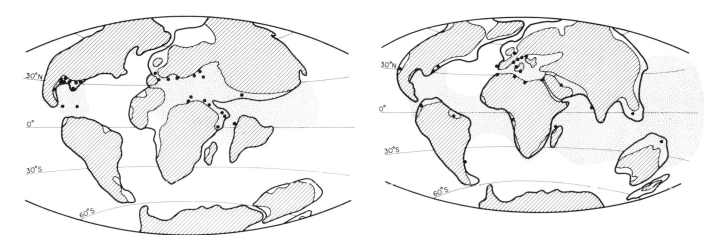

Maps showing the distribution of seagrasses (stippled) and sirenians (black circles) in Eocene (left) and Miocene times. Seagrasses are a major food of seacows, hence their parallel distribution in tropical waters about 30° north and south of the equator. (Partly after Braiser).

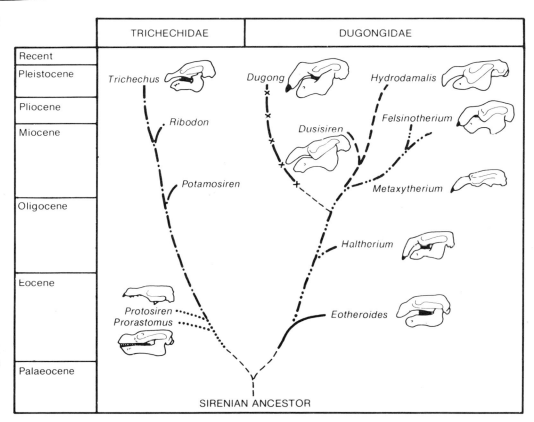

Skull of *Prorastomus* (27 cm long), a very primitive seacow from the Eocene of Jamaica. The thickened bony snout and bilophodont teeth suggest sirenian affinity; presence of five premolars is indicative of an origin from very early and primitive mammals.

	TRICHECHIDAE	DUGONGIDAE
Recent		
Pleistocene		
Pliocene		
Miocene		
Oligocene		
Eocene		
Palaeocene		

SIRENIAN ANCESTOR

Relationships of sirenian genera. The tusks and cheek teeth are in black.

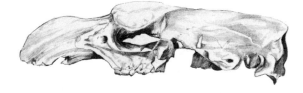

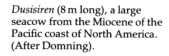
Dusisiren (8 m long), a large seacow from the Miocene of the Pacific coast of North America. (After Domning).

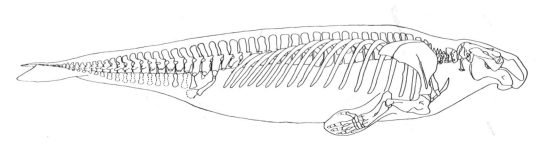

'Seahorses'

If any animals deserve the name 'seahorse', it should go to the desmostylians, extinct pony-sized mammals that lived in the coastal waters of the northern Pacific during the Miocene.

In the 1960s, during the course of excavations at Stanford University (California) for the foundations of a linear accelerator, a complete skeleton of a desmostylian was discovered; it received the name *Paleoparadoxia* (fossil puzzle). The body was pony-sized, horse-like and lacked the pachyostosis so ubiquitous among the seacows. The limbs were short and stout, the feet large and wide, and each had four hoof-like nails. On the fore limb the ulna and radius were fused (ankylosed), so the foot could not be turned without rotating the whole leg. *Paleoparadoxia* was probably amphibious, moving on land rather in the manner of sealions, while in the water it could have walked along the bottom of shallow coastal seas much as hippos walk over river and lake bottoms. The skull of *Paleoparadoxia* has overall similarity in shape to a horse, with deep mandible, elongated snout and a battery of cheek teeth, but there the similarity ceases. *Paleoparadoxia* has incisors and short tusks in both upper and lower jaws; these are separated by a gap (diastema) from the battery of cheek teeth that appear to have been replaced cyclically as in elephants. The structure of the desmostylian cheek teeth is unique among mammals; each tooth comprises a cluster of stout dentine tubes covered with thick enamel, and it is this feature that gives the order its name (desmos = chain, stylos = pillar).

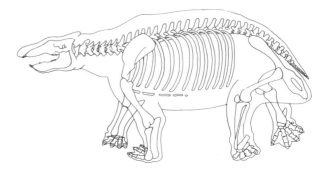

Skeleton of *Paleoparadoxia*; note the peculiar stance with inturned feet. (After Shikama).

There are only four known genera in the order, whose time range is earliest to latest Miocene. Fossils are found in coastal deposits of the Pacific Ocean between 25° and 53° north, on both west and east coasts. Japan, Kamchatka, Alaska, California and Mexico have all yielded desmostylians. There is a record from Florida, indicating passage through the Central American seaway between the Pacific and the Atlantic Oceans in mid Tertiary times.

The diet of desmostylians can only be guessed. Their limbs and short shovel-like tusked mouth would be consistent with a habit of walking on the sea bed and bottom feeding, using the tusks to prize off food, be it seaweed, seagrass or even mollusc. Coming out on land or into swamps to feed, they would parallel hippos;

Restoration of *Paleoparadoxia*, a desmostylian from the Miocene of California. These strange 'seahorses' probably led a walrus-like life.

feeding on molluscs they would parallel walruses. The late desmostylians and early walruses are found in the same deposits along the north Pacific coast of America; as the desmostylians became extinct the walruses diversified. The desmostylians still remain a group of unknown affinities, unknown origin and unknown feeding habits – all this despite our knowing complete skeletons.

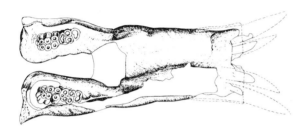

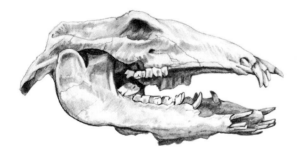

Skull and mandible of *Desmostylus* from the Miocene of the northern Pacific. The elongate snout has well-developed canine tusks in both jaws; the cheek teeth are made up of clusters of thickly enamelled tubules. (After Gregory and Reinhart).

Skull of *Paleoparadoxia* (length 40 cm) from the Miocene of Japan.

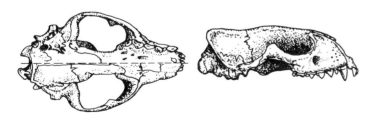

Skull of *Enaliarctos* (length 21 cm) in dorsal, ventral and lateral views. The cheek teeth are dog-like and not yet reduced to a single cusp. However, the restriction behind the large eyes is typical of sealions. (After Mitchell & Tedford).

Carnivores

Seals and sealions

To witness a troop of sealions perform in a circus or in a seaquarium is to experience a demonstration of command response rarely surpassed in animal training. Sealions have brains that are large in proportion to their size and like human brains have extremely complex surface folds. They can dive to depths of 200 metres and live in antarctic temperatures. They breed in very large colonies with females far outnumbering the males. Sealions and seals are undoubtedly highly successful as marine carnivores, and they have achieved that success in a mere 25 million years of evolution from otter-like and bear-like ancestors. The fossil record is now sufficiently good to infer that seals and sealions are not closely related; they can be derived from different families of terrestrial carnivores, but because both stocks have adapted to a broadly similar life, they have evolved many features in common. Walruses form a third living stock, related to sealions.

Otters swim using their long strong tails and their paddle-shaped webbed feet. The tail is reduced to a stump in almost all living marine carnivores; sealions use their front feet as the main swimming organ and seals derive the main thrust from their hind feet. To achieve maximum efficiency in transmitting power to the foot, the proximal part of the limb is short; to achieve maximum thrust, the foot is large; indeed the foot length may exceed that of the rest of the limb. The feet are webbed, so that when the digits are splayed out, a large surface area is presented. The hind feet are used as hydrofoils in a figure of eight movement that gives thrust on both the power and recovery strokes. On land, sealions can angle their hind feet to give them support and even some thrust. Seals, however, are unable to extend their hind limbs on land and rely on body movements to wriggle along with the fore limbs giving purchase while the trunk is drawn forward. Despite their seeming clumsiness on land, seals and sealions are surprisingly adept and can travel many kilometres over mountainous terrain.

Heat retention is a major problem for an aquatic mammal. The thermal conductivity of water is 25 times greater than that of air, so to overcome this, aquatic mammals grow large (thus achieving a small surface area to volume ratio) and they develop a thick layer of fat (blubber) to insulate them. To enable them to dive deep in search of food they have many specializations of their blood circulation, involving ability to cut down on oxygen consumption to the body, enlarged circulation to the brain, and enlarged venous sinuses; the latter can be traced in fossils through the enlarged vertebrarterial canals in the cervical vertebrae. The ear region has had to be modified to protect it from increased pressure (20 atmospheres at 200 metres) and to become receptive to water-borne sounds.

Dentitions become very simplified in fish eaters; basically all the teeth have a single cusp. Walruses have secondarily specialized to a molluscan diet; they have tusks for prizing off the shells from the sea floor and flattened cheek teeth to crush them. Sexual dimorphism is very marked in almost all living marine carnivores; elephant seal bulls can exceed 3·5 tonnes and they ferociously guard their vast harems.

ORIGIN AND EVOLUTION   Five families of marine carnivores are known – the extinct Enaliarctidae and Desmatophocidae, and the living Otariidae (sealions), Odobenidae (walruses) and Phocidae (seals). These are usually all grouped as Pinnipedia, a suborder of the Carnivora. However, with recent detailed study of the ear regions of aquatic carnivores, the view that phocoids (that is seals) and otaroids (the other four families) had separate (diphyletic) origins from terrestrial carnivores, has gained considerable support. Yet, while palaeontologists tend to see the pinnipeds as diphyletic neozoologists find increasing support for their single or monophyletic origin. The evidence comes from studies of chromosomes and serum albumins (types of protein). The chromosomes serum counts are clearly similar : $2n = 36$ otariids, 32 in odobenids, and 32 or 34 in phocids; 32 can be derived from 34 by fusion. Arnason remarked 'the chromosome morphological and banding homologies between otariids and phocids confirm the theory of monophyletic origin of th pinnipeds'. Sarich investigated the evolution of albumins in pinnipeds, using immunological distance as a measure of relationship. He showed that the bear *Ursus* was slightly closer to a sealion than to a seal, and that the sea otter *Enhydra* was slightly closer to a seal than to a sealion. However, seals and sealions were shown by Sarich to be closer to each other than to any non-pinniped family. The validity of phylogenetic deductions based on this karyological and serological evidence has been questioned and as yet many problems (variations, rates of change, convergence etc) are unresolved.

The family Enaliarctidae is founded on animals from the earliest Miocene of California; they were rather larger than the largest living otters but smaller than any living sealion. Some features point to their recent ancestry from amphicyonodontine ursids, a now extinct bear lineage; these include the overall shape of the skull, the pattern of the brain, the presence of carnassial teeth and crushing molars. *Enaliarctos* also has a number of features that point in the direction of sealions – large orbits (for large eyes), enlarged anterior nasal chamber, enlarged infraorbital foramen (for the nerve to the upper lip and whiskers), reduction of the ethmoturbinals and reduced olfactory lobes (and hence diminished sense of smell), enlarged posterior lacerate foramen (for improved drainage of blood from the brain during diving). With these characters we can envisage *Enaliarctos* as a coastal dweller perhaps not unlike the sea otter, but feeding on a variety of sea foods, including fishes. From enaliarctids can be derived, in late early Miocene, the desmatophocids, known from California, Oregon and Japan.

The only known genera of Desmatophocidae are *Desmatophocus* and *Allodesmus*, and they are in many characters intermediate between *Enaliarctos* and the otariid sealions. The dentition comprises homodont teeth, each with single cusp; in *Desmatophocus* each tooth has two roots, but in *Allodesmus* there is only one root as in otariids. The eyes are very big, even by sealion standards; probably the ear was not yet fully developed

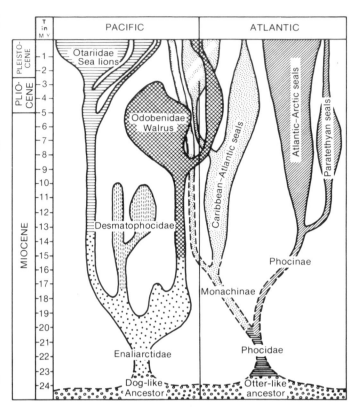

The relationships of aquatic carnivores. (After Repenning, Ray & Grigorescu).

Head and shoulders of a sealion (top, left), walrus (above), and seal.

The relationships of aquatic carnivores as deduced from serum albumins. (After Sarich).

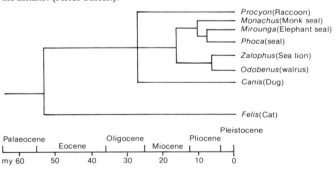

for underwater hearing and the eyes were still very important in hunting fish. Early desmatophocids overlapped with the last enaliarctids and the later desmatophocids overlapped with the first walruses and sealions.

Neotherium from the mid Miocene of California is the earliest known odobenid; it is little changed from *Enaliarctos*, but is larger and in limb bones shows evidence of sexual dimorphism. Then a little later in Miocene times, along the same coasts were populations of *Imagotaria*, a walrus-sized mammal with sealion-like dentitions and again displaying evidence of sexual dimorphism, suggesting it fed in the open sea and may have come back to a rookery to breed. By late Miocene and early Pliocene times there were at least five genera of walruses on the north Pacific coast; many were rather sealion-like. Others such as *Aivukus* had broad cheek teeth, large upper canine tusk and deep mandible with small canine; they could crush food and were probably bottom feeders in shallow water, as are modern walruses. Some of the molluscivorous walruses migrated through the Central American seaway in late Miocene times and spread north along the Atlantic coast of America and reached European waters by early Pliocene. Their lineage, however, became extinct in the Pacific during the Pliocene and in late Pleistocene times the walruses migrated around the Arctic Ocean back into the north Pacific.

The otariids today comprise six genera (sealions, eared seals and fur seals). The earliest known form is *Pithanotaria* from the mid Miocene of California. It was

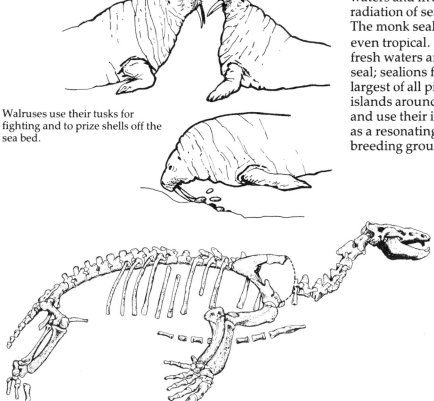

Walruses use their tusks for fighting and to prize shells off the sea bed.

very small; about the size of a sea otter and shows no signs of sexual dimorphism. The cheek teeth were homodont and had double roots; this last feature is one of the few that distinguish it morphologically from a living fur seal. In early Pliocene times about five million years ago otariids began to spread into the southern hemisphere where remains have been found along the coast of Peru. By late Pliocene times they were replacing early walruses in the northern hemisphere and spreading around Antarctic waters. Today the southern fur seal *Arctocephalus* is found in South America, South Africa, Australia, New Zealand, South Georgia, and in the tropics of the Galapagos Islands; it also has one surviving outpost in the northern hemisphere, along the California coast.

Summarizing, we can recognize three great waves of otaroid diversification to marine life. First the desmatophocids 16–14 million years ago in the north Pacific; second the odobenids 8–5 million years ago, again in the north Pacific, and finally the otariids in the last three million years in both north and south Pacific and in polar waters.

Although phocid seals are more abundant today than otariids, their fossil record is not good. In early Miocene times in central France were freshwater lakes with abundant otter-like mammals named *Potamotherium*; in many ways this animal was more adapted to aquatic life than living otters and may already have been a seal, depending on where and how the dividing line is drawn. Before the Central American seaway closed seals invaded the Pacific, spread south along the coasts of Peru and crossed to Hawaii, while others adapted to cold waters and invaded Arctic and Antarctic seas. The radiation of seals is rather greater than that of sealions. The monk seals frequent temperate waters and some are even tropical. Seals will go far up rivers and even live in fresh waters and in inland seas as does the Lake Baikal seal; sealions frequent only open oceanic waters. The largest of all pinnipeds, the elephant seals, are found on islands around Antarctica. Bulls may weigh 3·5 tonnes and use their inflatable proboscis or elephant-like trunk as a resonating chamber when trumpeting on the breeding grounds to defend their harem.

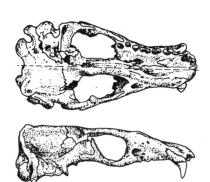

Skull of *Allodesmus* (length 38 cm) in dorsal, ventral and lateral view; the canine is well developed and the cheek teeth are all single rooted. (After Barnes).

Skeleton of the Miocene sealion *Allodesmus* (length 2 m) from the Pacific coast of North America. (After Mitchell).

Restoration of *Enaliarctos*, an ancestral sealion from the early Miocene
of the Pacific coast of USA. This small animal was probably not fully
aquatic; the large eyes and nasal openings are, however, quite sealion-
like.

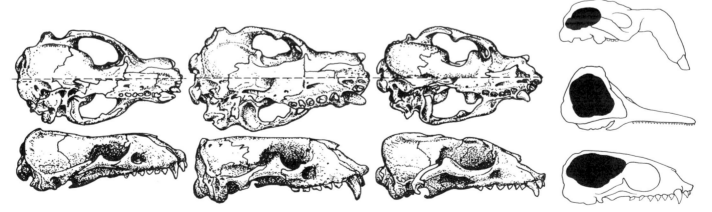

Skulls of two walrus genera, *Imagotaria* and *Aivukus*, and the sealion
Thalassoleon, all from the Miocene of the Pacific coast of USA. *Imagotaria*
is a very primitive walrus, *Aivukus* is more advanced with larger tusks
and broad crushing cheek teeth. *Thalassoleon* has the simple dentition
of a sealion. (After Repenning and Tedford).

The skulls of a dugong (top),
whale (middle) and sealion
(bottom); all reduced to the same
length to show the proportional
size of the brain (in black).

Aivukus, a shell-eating walrus
from the Miocene of the northern
Pacific.

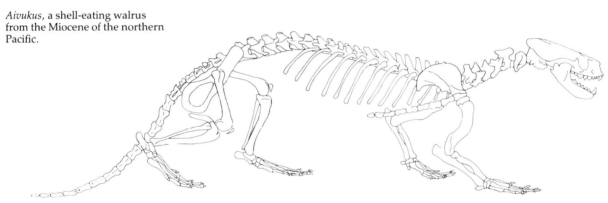

Skeleton and restoration of *Potamotherium* from the early Miocene of
France. This animal is either a semi-aquatic mustelid or a very early
seal.

Whales and dolphins

Whales include the largest animals that have ever lived; blue whales can reach 30 metres in length and 130 tonnes in weight. Only the need of air for respiration prevents whales from remaining permanently submerged in the oceans; they have otherwise completely adapted to aquatic life, with all the versatility of fishes. The order Cetacea comprises 44 living genera and about 150 fossil genera. Of all the mammals they are the most specialized; everything about their anatomy and physiology is orientated toward success in an aquatic environment. They are found in oceans worldwide, and freshwater whales are found in rivers in South America, India and China. The body form is streamlined, the tail with its horizontal two-fin fluke is the main organ of propulsion, the hind limb is absent, the fore limb is short, totally webbed as a flipper and used for steering and roll control. Whales have a thick fat layer (blubber) to help them maintain body heat; they have no sweat glands, but a tear gland persists to protect the eye from salt. Their sense of smell is lacking and their vision is poor, however, their sense of touch is acute and their underwater hearing is exceptionally well developed. Their brains are large and complex; though never proportionately as large as in man, *Tursiops* (bottle nosed dolphin) has a brain of about the same size as that of man, but the whale weighs about twice as much as a man, hence the brain/body weight ratio is only half that of man. All whales are basically carnivores, feeding either on squid or fish, or on shoals of plankton. Whales can stay submerged for over an hour and reach great depths; sperm whales have been recorded at depths exceeding 1000 metres. While whales do still breathe air like land mammals, they do so with much greater efficiency; the tidal volume of air passing through their lungs is ten times greater than in man (8·8 litres per kg of body weight in whales, and only 0·8 litres per kg in man). Whales can utilize 10 per cent of the oxygen in the air, while man utilizes only 4 per cent.

The bones of whales are filled with oil which aids flotation and acts as an energy reserve, especially during migration. The snout is elongate and the rest of the skull telescoped; so much so that the maxillae on the snout may touch the supraoccipitals at the back of the brain case. The cervical vertebrae are short and the middle ones are often fused. There is no sacrum, no hind limb and only a floating remnant of the pelvic bone. The pectoral girdle lacks a clavicle, the humerus is short and the paddle comprises four or five digits, each with many phalanges, all enclosed in a web of skin which is highly vascularized and can be used for heat control.

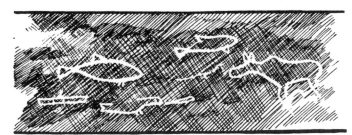

The oldest drawing of a whale; a Neolithic rock drawing from Rødd⊘y, Norway.

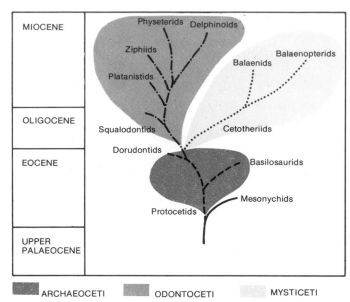

Relationships of cetacean suborders and families.

ORIGIN AND EVOLUTION   The order Cetacea contains three suborders – the extinct Archaeoceti, which include the earliest known whales and probably the ancestors of the other two stocks, the Odontoceti or toothed whales and the Mysticeti or baleen whales. The earliest records of archaeocetes come from the early Eocene of Pakistan. The best preserved of three genera there is the partial cranium of *Pakicetus*; it has upper cheek teeth with three cusps arranged in a triangle as in mesonychids and has lower teeth with some resemblance to *Protocetus* from the mid Eocene of Tethys. The ear of *Pakicetus* would have been most efficient in air suggesting that the animal had not totally become land free, as are the

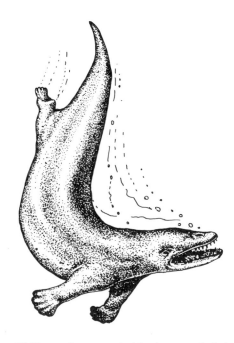

Restoration of *Pakicetus*, the most primitive known whale from the early Eocene of Pakistan.

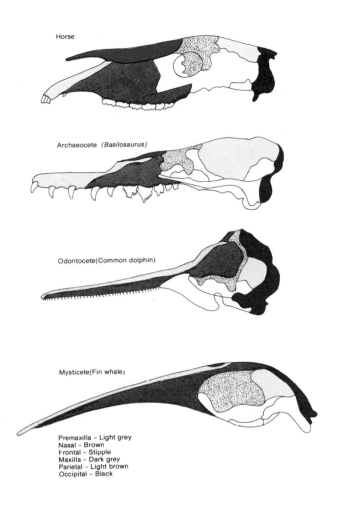

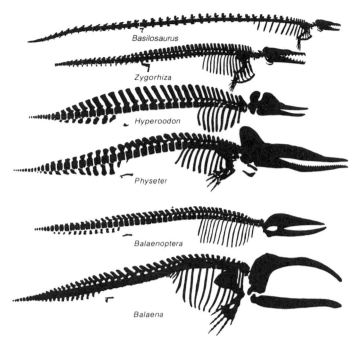

Silhouettes of two archaeocetes (top), two odontocetes (middle) and two mysticetes (bottom). Note the change in the proportion of the skull size. (After van Beneden, Gervais and Kellogg).

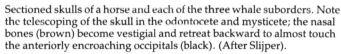

Sectioned skulls of a horse and each of the three whale suborders. Note the telescoping of the skull in the odontocete and mysticete; the nasal bones (brown) become vestigial and retreat backward to almost touch the anteriorly encroaching occipitals (black). (After Slijper).

Cheek teeth of a zeuglodont archaeocete (left) with a cascade of cusplets either side of the central cusp and the shark-like tooth of a squalodontid odontocete (right).

present-day sealions. Several archaeocetes are well known from the mid Eocene of Egypt and a few have been found in Virginia and Texas.

By late Eocene times the whales had become more widespread, being recorded from boreal seas and also from the southern hemisphere (New Zealand and Antarctica). Three basic types of archaeocete are recognized. The small primitive protocetids (e.g. *Protocetus*); the body reached about 2·5 metres and the skull about 60 cm with an elongate snout for catching fish. The medium-sized dorudontids (e.g. *Zygorhiza*); with a body around 7 metres and a skull about 90 cm long were very like porpoises. The third group were the large long-bodied zeuglodonts (e.g. *Basilosaurus*); in these the body could exceed 20 metres in length but the skull was only about 1·5 metres long. Thus, they were very different from a 20-metre sperm whale where the skull would make up 30 per cent of the total length, while in the baleen whales the skull may account for over 40 per cent of the total length.

The archaeocete skull retained traces of terrestrial ancestry, particularly in the brain case and in the retention of an olfactory sense. The snout, however, was elongate, with a heterodont dentition which usually exceeded the unreduced placental maximum of 44 teeth. The anterior teeth became progressively enlarged and incisiform to capture prey, and simultaneously the posterior premolar and molar teeth were progressively reduced. The molars still retained two or three roots, with a cascade of cusplets anterior and posterior to the high central cusp (i.e. zeuglodont). The ear was coupled acoustically with seawater and probably functioned in underwater intraspecific communication, and possibly in food detection, though not in high frequency reception. In the absence of sonar, vision was probably the major sense in archaeocetes. The fore limbs were flat and paddle-like and the hind limbs vestigial or absent. The powerful swimming tail was flexed dorso-ventrally and these whales may have been capable of coming ashore, much as seals do, for parturition and copulation. Archaeocetes were piscivorous dwellers in shallow offshore seas, though the larger zeuglodonts probably hunted in deeper and more open waters.

The odontocete or toothed whales probably arose from the dorudonts in late Eocene times. They diversified very rapidly and over 100 fossil genera are now recorded. Most odontocetes are dolphin-sized, though one family – the Physeteridae or sperm whales – has since mid Miocene times had genera which reached 12 metres in length. There are two dental trends prominent among odontocetes; in one the tendency to increase the number far above the normal 44 maximum for mammals, a trend which has its acme in some porpoises which have over 300 teeth. These polydont dentitions are usually made up of simple single rooted and single cusped teeth, though in the earlier taxa two roots are usually to be found. The second trend is to reduce and even lose all the teeth, as witnessed amongst most of the ziphiids, the beaked and bottle nosed whales. Squid and fish are the major items of diet among toothed whales. As with land carnivores, speed is essential in capturing their prey; the fastest dolphins can achieve almost 50 kh^{-1}. Their bodies are highly streamlined and their tail musculature produces very powerful thrusts.

A further requirement is a well-developed communications system, and in this the toothed whales probably surpass all mammals other than man. In early odontocetes, such as *Agorophius* from the Oligocene of North America, a hint of skull telescoping is seen and this becomes progressively more dominant. By Miocene times the maxillae bones reach back to make contact with the occipitals. The single nasal opening is on the roof of the skull over the laterally expanded brain case. This telescoping is made possible through the loss of many of the functions that are required of a terrestrial mammal. The neck has shortened, allowing reduction of the neck muscles and consequently reduction of the occipital region whose almost sole function is to protect the posterior half of the brain. The dietary habits do not necessitate chewing, so all jaw musculature can become greatly reduced, and thus much of the skull is lighter and bones can be more lightly built. The loss of an olfactory sense and posterior shift of the nasal opening enables the anterior snout region to become slender, housing only a row of simple teeth.

From archaeocetes of the late Eocene and early Oligocene, there developed the squalodontids, so named on account of their shark-like triangular cheek teeth; they were abundant and widely successful in temperate waters and in many ways probably resembled modern dolphins. But as we trace the evolution of odontocetes through mid Tertiary times there appear other skull adaptations not found among the now extinct squalodontids. These are the features associated with the development of echolocation. The bone forming the main part of the auditory capsule (the periotic) becomes uncoupled from the skull and remains attached only by soft tissue; the tympanic bulla becomes extremely dense and fuses with the periotic. The skull acquires a bilateral asymmetry when viewed dorsally and the nasal passages are asymmetrical. The facial region becomes basined to accommodate the melon, a fatty cushion, which is seen in its extreme form in the sperm whale. Asymmetry is first evident in mid Miocene odontocetes belonging to three different families, the ziphiids, the delphinids and the physeterids, and is recorded from Europe, North and South America. This suggests that asymmetry may have evolved independently in each lineage. The degree of asymmetry does not correlate with size, age or lineage.

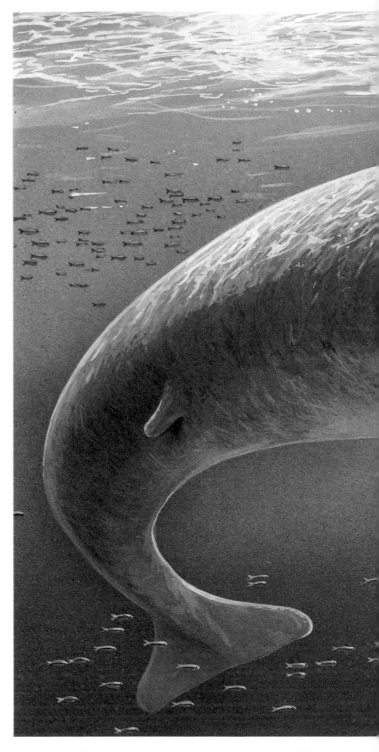

Restoration of *Protocetus*, a small archaeocete whale (length 2·5 m) from the Eocene of the Mediterranean.

While it is found among all living families of toothed whales, there are species in which it is almost absent (e.g. Brazilian river dolphin *Stenodelphis* and the killer whale *Orcinus*). The asymmetry is more an indication of adaptation to catching high speed prey, to navigating on long migrations and for communications to keep the school together.

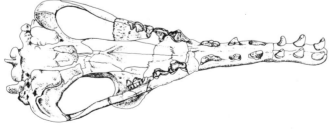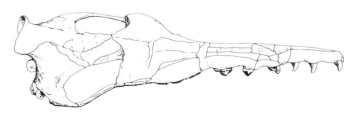

Skull of *Protocetus* (length 50 cm) with elongate snout and pointed teeth for holding fish. (After Fraas).

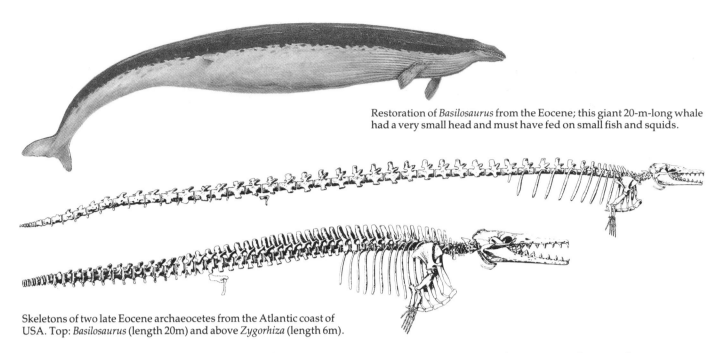

Restoration of *Basilosaurus* from the Eocene; this giant 20-m-long whale had a very small head and must have fed on small fish and squids.

Skeletons of two late Eocene archaeocetes from the Atlantic coast of USA. Top: *Basilosaurus* (length 20m) and above *Zygorhiza* (length 6m).

The mysticete or baleen whales are less diverse taxonomically than the toothed whales. However, they include the largest animals that have ever lived, with the great blue whales sometimes exceeding 130 tonnes. Mysticetes are toothless, but the jaws are very large and usually arched outward to accommodate the plates of baleen or whalebone, which are sheets of toughened skin with a dense hairy fringe. The whales feed mainly on plankton which is sieved through the baleen plates, caught in the hair and sucked off by the tongue.

Mysticetes appear to have arisen from archaeocetes and the earliest known form is *Mauicetus* from the mid Oligocene of New Zealand. It was already toothless and probably provided with baleen plates, which do not fossilize. The skull shows primitive telescoping, but the changes are different from those seen in odontocetes, and no asymmetry is ever seen in any baleen whales. However, they do have an echolocation system, though it is less complex than in the toothed whales. The evolution of the baleen whales correlates closely with the

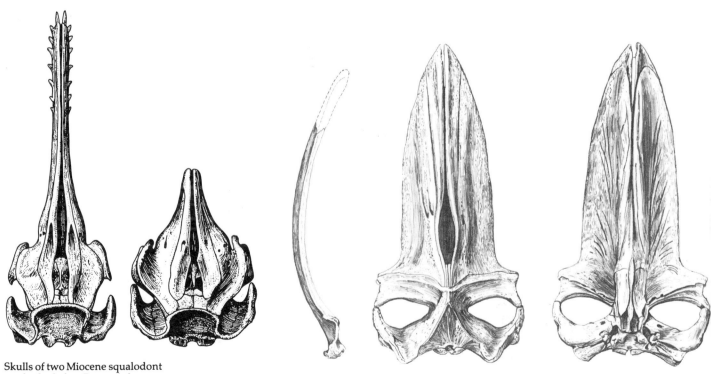

Skulls of two Miocene squalodont toothed whales, *Prosqualodon* from Patagonia and *Squalodon* from eastern USA. Note the absence of asymmetry in the skulls. (After Kellogg).

Skull of *Pelocetus* (length 65 cm), a cetotheriid baleen whale from the Miocene of North America. The wide rostrum housed the baleen plates and the mandible (top) is strongly curved. (After Kellogg).

initiation of the circum-Antarctic current, which became fully established by the mid Oligocene and with the overall cooling of ocean waters in Oligocene times. These environmental changes appear to have triggered a dramatic increase in plankton productivity. The baleen whales were rapidly able to exploit the potential of these planktonic blooms. During the Oligocene and Miocene the cetotheriid baleen whales became very abundant and diverse in oceans around the world. Today many baleen whales are on the verge of extinction through man's over-exploitation.

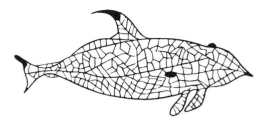

Norwegian rock drawing of a dolphin: original 2 m long.

Restoration of *Eurhinodelphis* (2 m long), a long-snouted porpoise from the Miocene of the north Pacific. The tip of the snout is without teeth, but the rest of the jaws are well armed with fish-holding teeth.

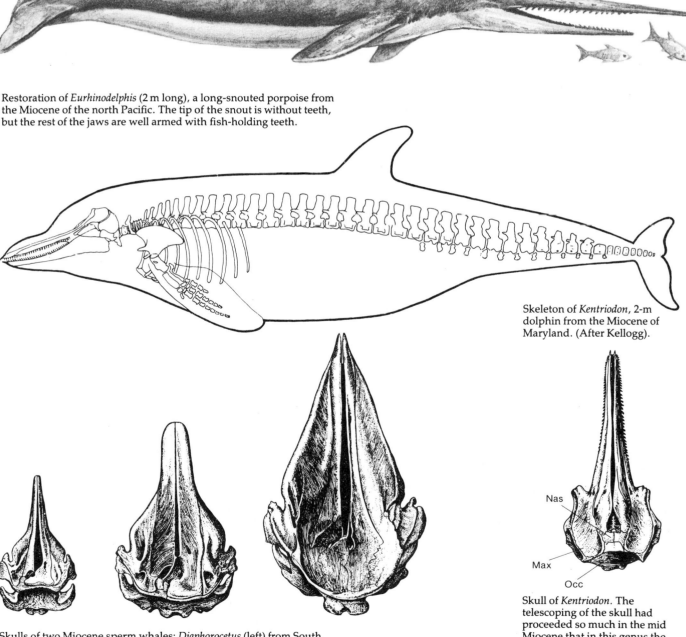

Skeleton of *Kentriodon*, 2-m dolphin from the Miocene of Maryland. (After Kellogg).

Skulls of two Miocene sperm whales: *Diaphorocetus* (left) from South America, *Aulophyseter* (middle) from California and *Physeter* (right) the living sperm whale.

Nas
Max
Occ

Skull of *Kentriodon*. The telescoping of the skull had proceeded so much in the mid Miocene that in this genus the nasal bones almost touch the occipitals.

CHAPTER 8
Gliders and fliers

Aristotle in his book *Meteorologia* viewed the world as being made up of four elements – fire, air, earth and water. These are distinguished by properties – heat, cold, dryness and moistness. The organic world inhabits air, earth and water, but is destroyed by fire. Mammals, being air breathers, must have access to air, but cannot live exclusively in that element. The great majority of mammals are to be found at the interface of air/earth and there are a sizeable number to be found at the triple interface of air/earth/water – these include otters, seals and other semi-aquatic mammals. A very few species have the ability to maintain themselves largely within one element; such are the whales in the water, the moles in the earth and the bats in the air.

The ancient writers thought of bats as peculiar birds. Pliny called them winged mice and then went on to say they were the only birds which had teeth, which brought forth their young alive and gave them milk. The adaptation to aerial life was seen as overriding all other characteristics. There are birds which cannot fly but no bats which do not fly.

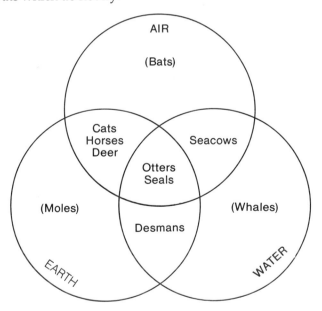

While all mammals breathe air, they spend differing amounts of their time on land or underground, on water or submerged, or in the air.

Moving through air

Parachuting
Animals can move through the air in three ways; they can parachute, glide or fly. In parachuting the only control is the rate of descent. Weight acts vertically downward and drag (the force developed parallel and opposite to the air flow) acts vertically upward. The rate of descent, dependent on weight (W) and surface area (S), is proportional to $\sqrt{\frac{W}{S}}$; thus the bigger the surface

area with respect to the weight, the slower the descent. Few if any mammals can be said to parachute in the true sense. Perhaps the nearest approach is in the free fall of, for example, a cat from a tree; in falling it spreads out its feet and tail to maximize its surface area and so reduce the acceleration of its fall.

Gliding
In gliding there is a horizontal component and the larger it is the longer the glide. A mammal jumping from a tree with a favourable updraught of air and with a high surface area to weight ratio may glide several hundred metres. Gliding is practised by mammals in three distinct orders – the marsupial gliding phalangers, the rodent gliding squirrels and the dermopterans or gliding 'lemurs'. None of these mammals has powered flight; they have folds of skin extending between the fore and hind limbs which greatly increase their surface area

A mammal has three options for aerial travel – parachuting, gliding or flapping flight.

Bat charm from ancient Egypt. The head of the mouse-tailed bat *Rhinopoma* was a symbol of wisdom. These bats have inhabited the Egyptian pyramids for thousands of years.

enabling them to remain airborne for long glides. It is probable that gliding evolved independently in each of these stocks though the only fossils known are dermopteran. (Among the reptiles there is a good fossil record of gliding lizards and also a living form, *Draco*.)

DERMOPTERA, THE 'FLYING LEMURS' The order Dermoptera is a very small order comprising the living genus *Cynocephalus* of south-east Asia and some fossil fragments from the early Caenozoic. The two species of *Cynocephalus* are in the vernacular misnamed 'flying lemurs'; they neither fly nor are they lemurs. They are arboreal squirrel-sized herbivorous mammals, living on fruit, flowers and leaves. They have large eyes and very

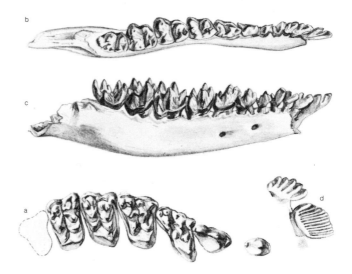

Plagiomene is found in the late Palaeocene and early Eocene of North America and is probably a dermopteran. The lower incisors are spatulate and forked, suggesting affinity with the comb-like incisors of *Cynocephalus* (d); *Plagiomene* (a–c); a) upper right cheek dentition b) and c) mandibular dentition (length 2.7 cm). (After Horsfall).

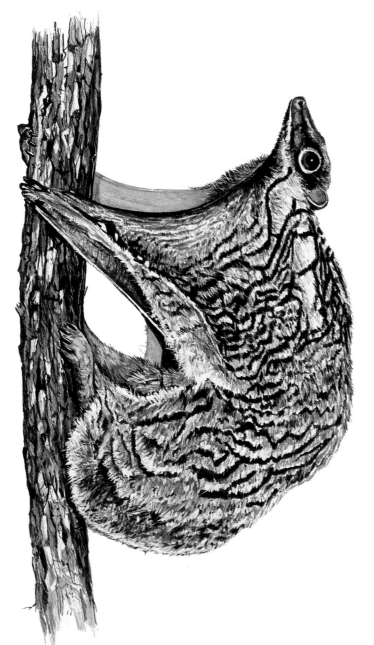

Cynocephalus, the colugo or 'gliding lemur' inhabits forested areas in south-east Asia and can glide over 100 m between trees.

Cynocephalus the colugo is a skilful climber and hangs in trees holding on with its long claws; weight 1–2 kg.

Skeleton of *Cynocephalus*; note the large eye sockets, the very long limbs and the long clawed digits for clinging to branches.

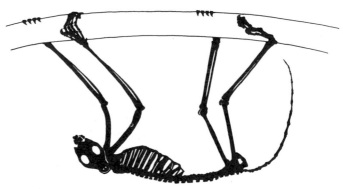

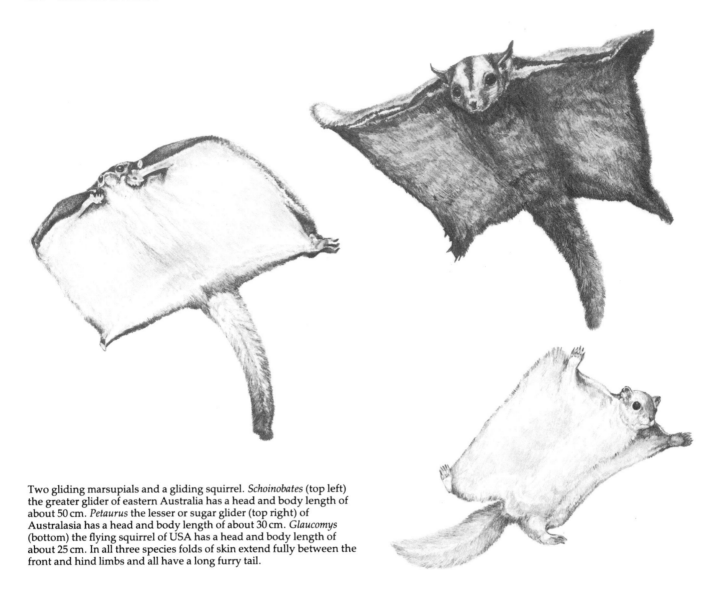

Two gliding marsupials and a gliding squirrel. *Schoinobates* (top left) the greater glider of eastern Australia has a head and body length of about 50 cm. *Petaurus* the lesser or sugar glider (top right) of Australasia has a head and body length of about 30 cm. *Glaucomys* (bottom) the flying squirrel of USA has a head and body length of about 25 cm. In all three species folds of skin extend fully between the front and hind limbs and all have a long furry tail.

long limbs, thus the skin folds which extend to the feet are very large. Skin folds also envelop the tail and the neck. *Cynocephalus* has been recorded gliding 136 metres between trees with a fall of 11 metres, giving a glide angle of only five degrees. One of the most distinctive features of *Cynocephalus* is the very specialized comb-like incisor teeth, though their function in feeding and grooming is poorly understood.

A few fossil jaws from the Palaeocene and Eocene of North America and some teeth from the early Eocene of France have been placed together in the family Plagiomenidae and referred to the order Dermoptera. The pattern of shearing facets on the molar teeth and the comb-like incisors are remarkably similar to those found in the living *Cynocephalus*. It seems likely that these herbivorous and arboreal mammals have nothing to do with the origin of bats. Recent immunological studies suggest that dermopterans have their closest relatives among the primates. The gliding specializations of the dermopterans are probably an evolutionary dead end and not an intermediate stage to true flight. The same applies to the other groups of gliding mammals, marsupials and rodents.

PETAURISTINAE, THE 'FLYING SQUIRRELS'   The gliding or 'flying' squirrels comprise a whole subfamily of squirrels, the Petauristinae, with about 12 genera and 36 species. They live mainly in the tropics of south and east Asia with one form in North and Central America – the American flying squirrel which has a Pleistocene fossil record. The gliding squirrels all live in forests, often at high altitudes. They have skin folds extending between the fore and hind limbs, leaving the tail free except for the base which is sometimes enclosed in the skin fold. In gliding the squirrels can use air currents to bank and the giant flying squirrel *Petaurista*, with a weight of over 1 kg, has been recorded in glides of 450 metres. The diet of these squirrels is usually fruit, nuts, leaves and insects, particularly larvae. None catches insects in flight.

PHALANGERS   The gliding phalangers of Australia comprise four genera and seven species. *Hemibelideus*, the brush-tipped ring-tail phalanger, has small skin folds less than 2 cm wide along the body sides. It lives in the rain forests of Queensland and jumps from tree to tree. The other three genera (*Schoinobates*, *Acrobates* and *Petaurus*) all have well-developed skin folds between the fore and hind limbs, but not including the tail; all are arboreal forest dwellers and feed on flowers, sap, leaves, insects and their larvae. Glides of over 100 metres are recorded at 30 degree glide angles.

Flying

True powered flight has been achieved among the vertebrates at least three times – in the reptiles with pterosaurs, in the birds and in the mammals with bats. In each case there are similarities and differences. The same aerodynamic principles apply to all organisms; all vertebrates utilize the fore limbs, but modify them for flight in different ways as seen in the feathered wing of the bird, the pterosaur wing supported by one elongated digit and the bat wing with a ribbed support of four digits like an umbrella. The bat wing works on the principle of the aerofoil to generate lift when air passes across it. In cross-section the wing is asymmetrical with

Sketches to show the sequential positions of the bat wing during flight. (After Pennycuick).

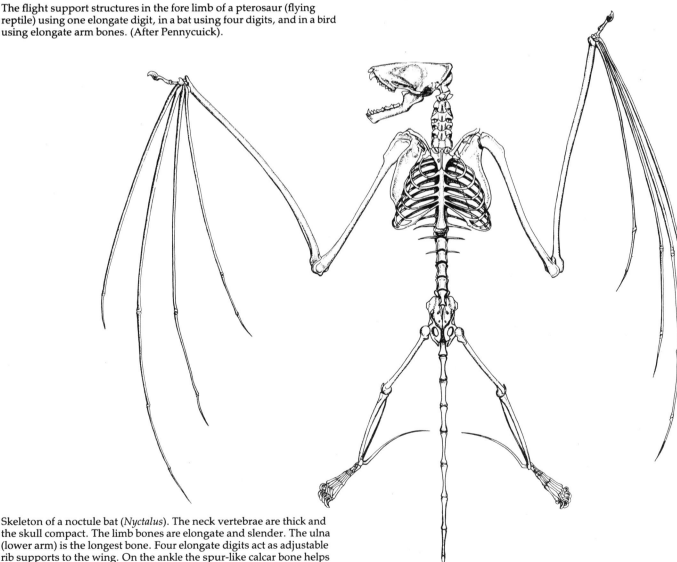

The flight support structures in the fore limb of a pterosaur (flying reptile) using one elongate digit, in a bat using four digits, and in a bird using elongate arm bones. (After Pennycuick).

Skeleton of a noctule bat (*Nyctalus*). The neck vertebrae are thick and the skull compact. The limb bones are elongate and slender. The ulna (lower arm) is the longest bone. Four elongate digits act as adjustable rib supports to the wing. On the ankle the spur-like calcar bone helps support the tail membrane.

a variable camber; the upper profile is convex and air flowing over the top has further to travel than air flowing underneath and must therefore flow faster to reach the posterior edge at the same time. In this way the downward pressure exerted on the top of the wing is less than the upward pressure of the air below, and this pressure difference provides the lift to keep the bat airborne.

Since air must flow over the wing for the aerofoil to generate lift there must be a means of moving the wing through the air and this movement produces a certain amount of resistance known as drag. There is parasitic drag due to the body, profile drag due to the wings and induced drag associated with lift generation. A good aerofoil will maximize the lift generated while minimizing the drag; this lift/drag ratio is a measure of aerodynamic efficiency. Bats and birds are quite similar in this respect – the ratio for a fruit bat is 6·8 which is slightly better than the figure of 6·0 for a pigeon.

The ribbing on the digits of the bat wing produces bumps on the upper membrane surface; the bumps have the effect of breaking up the boundary layer of air and so produce turbulence, which is very beneficial for slow flight. Birds can glide fast by folding back their feathered wings; bats cannot fold their membranous wings so far and hence can only glide slowly.

ECHOLOCATION Echolocation is used among mammals in three different orders – insectivores, whales and bats, and is used by some birds. Echolocation involves emitting high frequency sounds that are reflected off objects, then picked up by the animal and the pattern of impulses interpreted. The principle is used in radar where the signal received is imaged on a visual screen, or in sonar depth-finding where the signal is shown graphically. Most bats use frequencies in the range 20–80 kilo-Hertz; only a few bats use frequencies less than 20 kHz which is the upper limit of human hearing. The sound is produced by bats in the larynx which is structurally modified for the purpose, and emitted through the mouth or nose. The returning signals are picked up by the ears which usually have large pinnae and often a tragus or 'antenna' to aid precise direction determination of the echoes. Each bat species appears to have its own range of sounds, used for a variety of information gathering and it is possible to identify bats by their 'signature tunes'. Echolocation has necessitated major modifications of the nose and external ear which do not fossilize. However, it has also necessitated the development of specialized interpretation centres in the brain. The bat brain is characterized by highly developed inferior quadrigemina in the midbrain, an area known to be associated with auditory reception.

The evolution of echolocation is a necessary requirement for any animal that attempts to fly in the dark, feed on the wing and roost in caves. Some shrews are known to use echolocation in foraging and although we do not know the ancestral history of bats, it is highly probable that they evolved from a shrew-like insectivore. It is reasonable to think that an echolocation faculty may well have been present in the pre-bat ancestor.

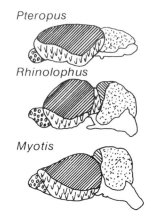

Side view of bat brains; top the fruit bat *Pteropus*, middle the horseshoe bat *Rhinolophus* and bottom the mouse-eared bat *Myotis*. All brains reduced to same overall length. Note the large neocortex in the fruit bat and the large cerebellum in the other two bats. (After Schneider).

Neocortex
Olfactory lobe
Cerebellum
Palaeopallium
Tectum

Bats

Living bats fall clearly into two groups – the Microchiroptera or microbats and the Megachiroptera or megabats. The microbats are by far the most abundant, with four superfamilies, 17 families, some 140 genera and about 780 species; they are almost worldwide in distribution, small, nocturnal and usually have small eyes, large ears and often specialized nasal outgrowths, the latter two characters associated with their universal use of echolocation. Their dentition usually has upper molars in which the cusps have shearing ridges arranged in a W-pattern. The great majority are insect feeders, some specializing on specific insects. The vampires include blood and nectar drinkers; several bats feed on fish and some even eat other bats. In most cases the food is taken on the wing.

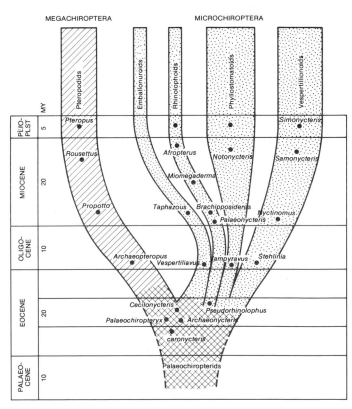

Family tree of bat superfamilies with the main fossil genera shown.

The second group comprises the megabats, the large fruit eating bats of the Old World tropics; there is only one family of these with about 40 genera and 170 species. They have very fox-like faces, hence one of their alternative names, the 'flying foxes'. Only one genus (*Rousettus*) is known to echolocate, using very low frequencies, the series of clicks being produced by the tongue. Fruit bats lack the big ears and nasal outgrowths of the microbats; their dentition is often very reduced since soft fruit makes up a very large part of their diet.

Fossil species

Marlstones of the Green River formation in Wyoming have yielded a rich fauna of early Eocene age comprising lake-living fishes, turtles and crocodiles along with snakes, birds and mammals. Occasionally fish are found that are almost complete and so well preserved that most of the scales are intact. The lake ecology appears to have been humid and subtropical. The most exceptional mammal found here is an almost complete articulated skeleton of a bat *Icaronycteris*; the animal was about the same size and proportions as the mouse-eared bat *Myotis myotis*. Although so similar to a living bat, *Icaronycteris* does retain a large number of primitive characters – the dentition comprises a large number of teeth (38), the molars are essentially insectivore-like, the sternal elements are unfused, there is no keel on the sternum, the tail is long and free, the radius is short, there is a claw on both the first and second digit and there is no calcar (a cartilaginous rod on the hind foot of living bats to support the wing membrane).

No estimate has been made of the weight of *Icaronycteris* and hence the wing loading cannot be calculated, but Jepsen has estimated that the aspect ratio of the wings (that is the wing length/mean wing breadth) is very low at around 2·8; the figure for *Myotis* is 6·7 and among birds a low figure would be 4·9 for the warbler *Dendroica*. Low aspect ratio implies very low flying speed.

Although *Icaronycteris* is the earliest known bat, it is so fully a bat that it tells us nothing about the origin of flight and very little about the ancestry of bats. *Icaronycteris* can readily be grouped with the microbats – indeed the first possible megabat is not known to have appeared for another 15 million years. Mid Eocene sites in Western Germany (Messel) and Eastern Germany (Geiseltal) have some well-preserved bat skeletons; the Messel site near Darmstadt was once worked for brown coal which occurs with oil shales. The rich fauna of this former lake, which was fed by rivers from the neighbouring hills, includes sponges, snails, insects, fish, frogs, crocodiles, snakes, birds and many mammals, including bats. The vegetation of palm and fig trees suggests tropical or subtropical climates. In the coal some of the fossils are preserved completely, with skin, feathers and hair; such is the case with some of the bats. Probably the freshwater lake had poorly aerated still waters so that an injured bat falling into the lake would sink to the bottom; there the anaerobic conditions would decelerate the decay and the specimen could become buried before it disintegrated.

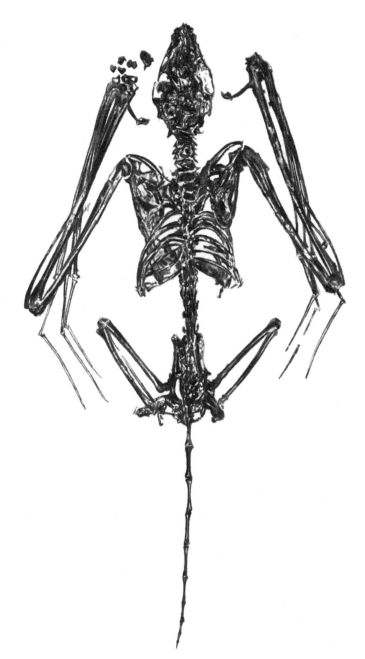

Skeleton of *Icaronycteris*, the earliest known bat, from the early Eocene of Wyoming. Overall length 12·5 cm. This animal was already a fully fledged bat and very similar to the living mouse-eared bat *Myotis*. (After Jepsen).

Head of *Artibeus*, a leaf-nosed fruit bat. These bats range from Mexico through Central America to northern Argentina. The elaborate leaf development on the nose is believed to be used to emit and direct sonar pulses. The large pointed ears with a tragus may aid echolocation of targets and are probably highly directional. The bats feed on a variety of fruit – figs, avocados, bananas, and mangoes.

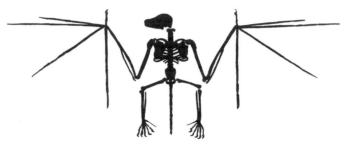

Restoration of the skeleton of *Palaeochiropteryx*, one of two bats from the mid Eocene of Messel in Germany. Head and body 7 cm. (After Revilliod).

Palaeochiropteryx preserved in oil shales formed in a freshwater lake at Messel. Like the living bat *Myotis* this Eocene fossil bat may have fed on the wing over the lake waters.

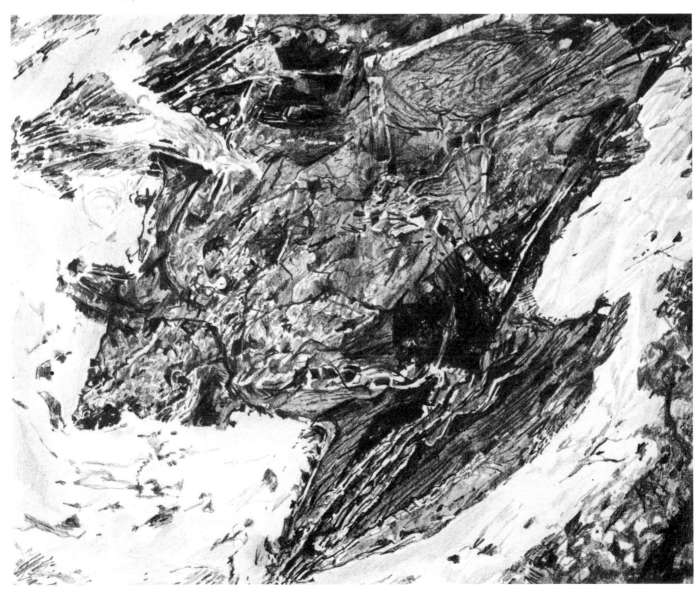

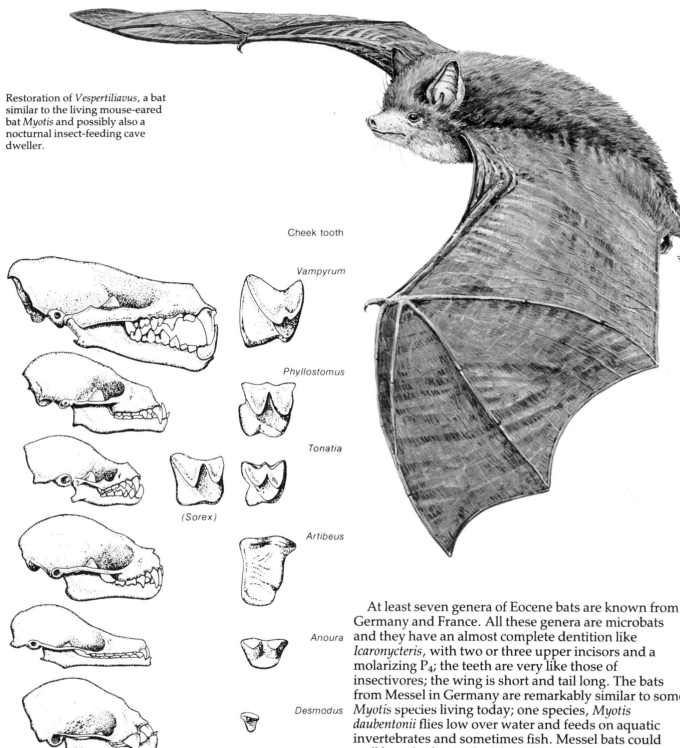

Restoration of *Vespertiliavus*, a bat similar to the living mouse-eared bat *Myotis* and possibly also a nocturnal insect-feeding cave dweller.

Cheek tooth

Vampyrum

Phyllostomus

Tonatia

(Sorex)

Artibeus

Anoura

Desmodus

Skulls of leaf-nosed bats (phyllostomids) from tropical America and the closely related vampire bat *Desmodus*. Their differing diets are reflected in their skull and dental adaptations. The insect-feeding *Tonatia* is the basic type with teeth similar to those of the shrew (*Sorex*). *Phyllostomus* is omnivorous and *Vampyrum* a carnivore with large canine teeth. The fruit-eating *Artibeus* has a short face and broad teeth to crush fruit. *Anoura* has an elongate snout and long tongue to suck nectar and the teeth are small. The blood-drinking vampire *Desmodus* has a short face with piercing incisors, large canine and vestigial cheek teeth. (After Yalden & Morris).

At least seven genera of Eocene bats are known from Germany and France. All these genera are microbats and they have an almost complete dentition like *Icaronycteris*, with two or three upper incisors and a molarizing P_4; the teeth are very like those of insectivores; the wing is short and tail long. The bats from Messel in Germany are remarkably similar to some *Myotis* species living today; one species, *Myotis daubentonii* flies low over water and feeds on aquatic invertebrates and sometimes fish. Messel bats could well have had a similar life style. Russell & Sigé grouped all these Eocene bats together into one family, the Palaeochiropterygidae, which is thought of broadly as probably including ancestors of the later microbats.

Fissure infillings in limestones of the Phosphorites du Quercy in France and at Egerkingen in Switzerland have yielded bat remains. Their age range is late Eocene to early Oligocene and bats represented include early members of the rhinolophoids (horseshoe bats), emballonuroids and vespertilionoids. These are three of the four superfamilies of microbats, the fourth is the

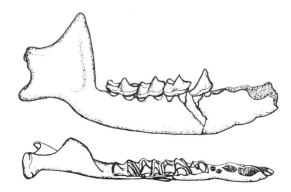

Mandible of *Vespertiliavus* (length 2·2 cm), from an early Oligocene cave deposit at Quercy in southern France. This bat may be the earliest cave dwelling bat, a habitat still utilized by its living descendants in the family Vespertilionidae. (After Revilliod).

Archaeopteropus skelton preserved in early Oligocene lignite of Venetia, Italy. This large bat with a 1 m wing span may have been the first megachiropteran or megabat. (After Dal Piaz).

exclusively American group the phyllostomoids, whose earliest known member is *Notonycteris* from the late Miocene of Columbia. The early Miocene tuffaceous sediments of western Kenya have some bat fragments which Butler has placed among the emballonuroids and rhinolophoids. The earliest African bat is an early Oligocene genus *Vampyravus* from the Fayum of Egypt; although its affinites are uncertain, it has a humerus which is very similar to that of the South American vampire bat.

Another fascinating and well-preserved Palaeogene bat is *Archaeopteropus* from the early Oligocene of Venetia, Italy. This bat could be considered to be the earliest megabat; however, although preserved in lignite, it is badly crushed and difficult to interpret. It is a large bat with a wing span of about one metre and with a long tail as in the living fruit bat *Notopterus*. Probably the only other Tertiary megabat is *Propotto* from the early Miocene of western Kenya; the jaw fragment was originally thought to belong to a lorisid primate, but is now considered to be a megabat (Walker, 1969).

Life style

Although bats are today second only to rodents in species diversity, there are still ten times as many birds as bats. Bats vary in size from shrew-like (2 g) to rabbit-size (1200 g), the largest bats having a wing span of 1·5 metres. Bats are crepuscular or nocturnal and mainly tropical, with only small numbers in the temperate regions. As night fliers catching food on the wing, bats far outnumber birds – this is the niche in which they have few serious competitors, the most effective being nightjars. However, even night flying and cave roosting do not enable bats totally to escape predators which include owls, hawks, hornbills, snakes and other bats. The birds catch the bats on the wing using their eyes in the half light. Snakes, particularly in caves, use infra-red sensors to detect their presence.

Bats tend to hibernate rather than migrate, and their use of caves as roosts has two geological consequences –

the accumulation of skeletal remains and accumulation of droppings to form guano deposits. Plio-Pleistocene caves around the world have yielded vast numbers of fossilized bat remains. Identification is often difficult without reasonably complete dentitions. However, almost all of these microbats have been ascribed to living genera; the bats do not appear to have suffered the late Pleistocene extinctions which greatly reduced the diversity of many mammalian orders.

Bat droppings are high in nitrates and have been worked commercially for fertilizers in several parts of the world. Bat guano from a cave on Mount Suswa in Kenya is due to colonies of *Otomops*, an insectivorous bat, and the packaged guano can be bought in Nairobi shops as a fertilizer. In New Mexico the Carlsbad limestone caverns have millions of the free-tailed bat *Tadarida brasiliensis*. The enormous guano deposits exceeded 30 metres in depth at the turn of the century. It is estimated that 100 000 tonnes of guano were mined during the first two decades of this century. During the American Civil War guano was used as a source of nitrate for explosives.

CHAPTER 9

Gnawers

Restoration of *Paracricetodon* from the early Oligocene of France; one of the earliest myomorph rodents and close to the ancestry of rats, mice, voles and lemmings.

The rodent model

To gnaw – to bite persistently – is the ubiquitous characteristic of the largest of all mammalian orders, the rodents: it is also seen in the small lagomorph order (rabbits and hares) and in a few extinct orders. All gnawing mammals possess large continuously growing incisor teeth. Unless they bite continually and wear their incisors down, they will be unable to close their jaws and soon die of starvation. So the gnawers most useful feature, the ability to bite almost anything from lead pipes to fully grown trees is also a burdensome liability. The worldwide success of the rodents, however, demonstrates the heavy weighting of the balance in their favour.

TRITYLODONTS, RODENT-LIKE REPTILES   During the late Triassic a family of mammal-like reptiles developed in a very rodent-like way; these were the tritylodonts and they persisted for some 50 million years into the mid Jurassic. They varied from mouse to rabbit in size, had enlarged incisor teeth followed by a long gap (the diastema), and a battery of multicusped cheek teeth, closely paralleling the rodents. They are found on all continents except Antarctica and Australia and probably filled an ecological niche similar to that occupied today by rodents.

MULTITUBERCULATES, RODENT-LIKE MAMMALS   The place of tritylodonts appears to have been taken during Jurassic times by an extinct order of mammals, the multituberculates (see p 42); they were in existence by the late Jurassic and may have an ancestry going back to the late Triassic. A varied and abundant stock in the northern hemisphere, they like tritylodonts had enlarged and sharp incisor teeth, followed by a diastema, and in the cheek a battery of multicusped teeth (hence the name of the order). The multituberculates took over the tritylodont ecological niche in the mid Jurassic and continued to thrive well into the Tertiary, only becoming extinct in late Eocene times. Ecologically they bridge the gap between the tritylodonts and the rodents. It is noticeable that the rodents only become abundant when the multituberculates declined. Thus, since the appearance of dominant seed-bearing vegetation in Triassic times, there has always been at least one stock of higher vertebrates which thrived by gnawing the hard plant tissues.

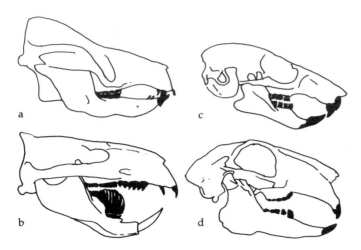

The skulls of four animals all adapted to gnawing and each belonging to a different order. Each has a pair of enlarged incisor teeth, followed by a diastema before the battery of grinding cheek teeth.

a. *Oligokyphus* from the early Jurassic, is an advanced mammal-like reptile belonging to the Tritylodontidae.

b. *Ptilodus* from the Palaeocene, is a multituberculate mammal, belonging to a stock that bridged the time gap between the tritylodonts and rodents.

c. *Muscardinus* a dormouse, representing the living rodents which extend back into the Palaeocene.

d. *Lepus*, a hare, representing the living lagomorphs which also extend back into the Palaeocene.

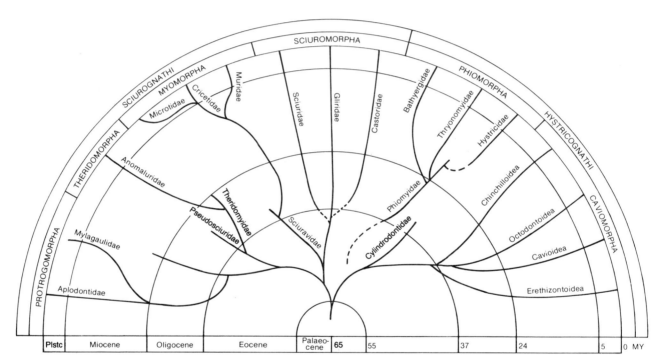

A phylogeny showing the probable relationships of the major stocks of fossil and living rodents.

Four different arrangements of the middle and deep layers of the masseter muscle in rodents.

 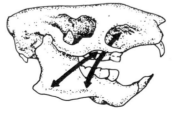 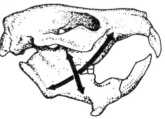 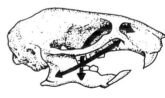

Protrogomorph: the primitive condition with the masseter attachments to the skull limited to the zygomatic arch area, e.g. *Paramys*, Palaeocene rodent.

Hystricomorph: middle masseter unspecialized, deep masseter passes through the infra-orbital foramen to attach anterior to the eye, e.g. *Hystrix*, porcupine.

Sciuromorph: deep masseter is unspecialized, middle masseter is attached anterior to the eye, e.g. *Sciurus*, squirrel.

Myomorph: middle masseter is attached anterior to the eye as in sciuromorph and the deep masseter passes up into the orbital area and through the infra-orbital foramen, e.g. *Mus*, mouse.

RODENTS AND LAGOMORPHS

Order	RODENTIA
Suborder	SCIUROGNATHI
Infraorder	Protrogomorpha
Family	†Paramyidae
	†Sciuravidae
	Aplodontidae
	†Mylagaulidae
	Sciuromorpha
	Sciuridae
	Gliridae
	Castoridae
	Theridomorpha
	†Pseudosciuridae
	†Theridomyidae
	Anomaluridae
	Myomorpha
	Cricetidae
	Microtidae
	Muridae
Suborder	HYSTRICOGNATHI
Infraorder	Phiomorpha
Family	†Phiomyidae
	Thryonomyidae
	Bathyergidae
	Hystricidae
Infraorder	Caviomorpha
Superfamily	Octodontoidea
	Cavioidea
	Chinchilloidea
	Erethizontoidea
Order	LAGOMORPHA
Family	†Eurymylidae
	Ochotonidae
	Leporidae

† = Extinct

MODERN RODENTS Living rodents such as squirrels, rats, mice and guinea pigs are grouped into 35 families, about 350 genera and around 1600 species, which is 40 per cent of all the known mammalian species alive today. Taking the fossil record into account adds another dozen families and about doubles the number of genera. Rodents have reached all continents except Antarctica; they inhabit an extremely wide range of habitats from cold tundra to tropical forest and from alpine heights to desert wastes. Most are terrestrial, small, scampering, quadrupeds with claws, long tails and long whiskers. A few occupy aquatic habitats (water rats and beavers), but there are no marine forms. Many are arboreal (such as the tree squirrels) and a few glide but none fly. Many South American forms are good runners and their nails are rather hoof-like. Bipedal jumping forms are common in deserts. The majority of rodents are mouse-to rat-sized; some, such as the harvest mice are amongst the smallest mammals (5 g). The largest living rodent is the capybara of South America which weighs around 50 kg. However, some of the fossil forms were much bigger, reaching bear and even rhinoceros size.

Beaver skull (*Castor*) to show the long teeth. The roots of both incisors and cheek teeth remain open and the teeth continue to grow throughout life.

Front view of the incisor teeth of *Bathyergus*, the mole rat. The large sharp upper incisors are heavily grooved. The animal lives in the sand dunes of Cape Province in South Africa and feeds on bulbs and fleshy roots.

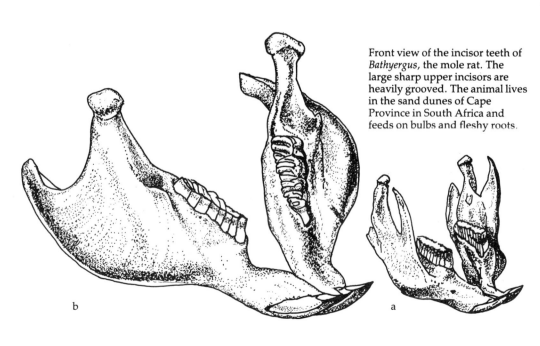

All known rodents, living and fossil, have one of two types of lower jaw.
a. Sciurognathous rodent with the deep bony surface directly below the cheek teeth for the insertion of the masseter muscle.
b. Hystricognathous rodent with an outwardly deflected bony flange for the insertion of the masseter muscle.

While the postcranial skeleton of rodents has not dramatically altered during the course of their evolution, the skull and dentition have become very greatly modified from the primitive condition – and it all relates to gnawing.

All rodents have only one pair of incisor teeth in the upper jaw and one pair in the lower jaw. The teeth are curved and continually growing so that as fast as they are worn away they are replaced by new growth. The incisors are triangular-shaped in cross-section with hard enamel present only on the front face; thus the area behind wears away more rapidly, ensuring that a sharp edge is maintained. Behind the incisor teeth is a diastema, while the back of the jaws are occupied by a battery of grinding teeth, usually four on each side of each jaw, P_4–M_3. A great variety of tooth patterns is developed; some, like squirrels and mice, have multicusped teeth reflecting a varied diet of tubers, berries, seeds, nuts, insects etc. Others such as voles, lemmings, pacas and capybaras have high prismatic teeth that are adapted to grazing sedges and grasses, and they have a digestive system to match. To grind the food procured by the gnawing, requires a substantial battery of teeth and also a powerful set of jaw muscles. The masseter muscle, which in closing the rodent jaw brings the mandible up and forward, is especially enlarged. The muscle inserts onto the deep bony flange of the mandible below and behind the teeth. In the most primitive rodents this bony plate lies in the same vertical plane as the incisor tooth; the condition is known as sciurognathous (literally squirrel-jawed). In slightly later and more derived rodents the bony plate is deflected outward, a condition known as hystricognathous (literally porcupine-jawed); this deflection allows elongation of the pterygoid muscle and may be related to propalinal chewing (moving jaws anteroposteriorly). All known rodents, fossil and living, manifest one or other of these two conditions.

The other end of the masseter muscle has its origins on the zygomatic arch; however, in rodents other than the primitive protrogomorphs, this is not sufficient anchorage, and the muscle origin extends forward to attach to bony areas in front of the eye. There are three major pathways in which this develops and these are the scuiromorph (squirrel pattern), the myomorph (mouse pattern) and the hystricomorph (porcupine pattern). On this basis many attempts have been made to accommodate all rodents within these four groupings; however, there are many problems, both morphological and phylogenetic.

Origin and evolution

Our knowledge of early forms is too slight to be able to assess the evolutionary history of rodents with any confidence and the arrangement adopted here of considering all rodents as falling within six major groups is as much a reflection of convenience as it is an expression of relationship.

Until a couple of decades ago the fossil record of rodents was very poor. It has been greatly improved upon in recent times by the use of screening techniques. Since rodent teeth are individually very small and make

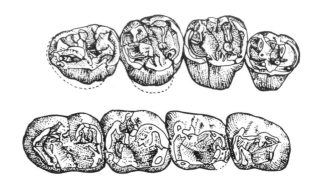

Cheek teeth of *Paramys*; top, the upper teeth (dP^4 – M^3) and bottom, the lower teeth (P_4 – M_3). The presence of three molars is a primitive feature and the simple four cusped pattern can still be discerned. (After Wood).

up such a small percentage of the total mass of a rock matrix, they are likely to be missed by field collectors. However, if the sand or clay is collected and washed through a fine seive the teeth will be quickly recognized in the concentrate. The yields are now sufficient for a stratigraphical zonation system to be erected for the Tertiary continental sequence using rodent faunas.

Rodent diversity

Protrogomorphs – the earliest true rodents

The earliest known fossil rodents are the paramyids of the late Palaeocene, which with their descendants in the Eocene are recorded from North America and Eurasia. Most paramyids were mouse-like scampering rodents, but some were as large as beavers. The nearest living descendant of this primitive group is *Aplodontia*, the mountain beaver of the Rockies. In early paramyids the incisor teeth are rounded in section with enamel extending around much of the tooth; as they evolved during the early Eocene later species of *Paramys* developed triangular-shaped incisor teeth with enamel restricted to the front edge. Changes can also be followed in the cheek teeth, with the cusps becoming more lophodont or ridged and the last molar developing into a fully functional part of the grinding battery. Paramyids are well known in North America in the early Tertiary, and less well known from Eurasia.

Two more extinct families of protrogomorphs deserve mention. The sciuravids were mouse- and rat-like animals; they are known from Eocene times in North America and Asia and were probably ancestral to the myomorphs, that is true mice and rats, of which more later. The mylagaulids were a very distinct family that lived in the Great Basin area of North America throughout the Miocene; they were about the size of beavers and lived in forested habitats. They had two extraordinary specializations: paired horns on their snout, and very long, strong compressed claws on their front feet. The specializations of the fore limb combined with their small eyes indicated an underground burrowing existence. But why should an animal that

Restoration of *Paramys* from the Eocene of North America. This primitive rodent was rather squirrel-like.

Skeleton of *Paramys* (length about 60 cm). Note the rodent-like skull, the long tail and the relatively long hind legs. (After Wood).

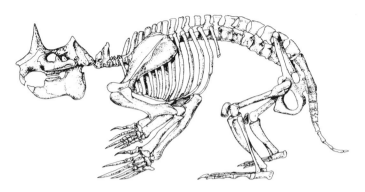

Skeleton of *Epigaulus* from the Miocene of the Great Basin, USA. The length of the head and body was about 30 cm; the skull has strong posterior crests for muscle attachment. The fore limbs are very strongly built with short stout bones and powerful muscles used in digging. (After Gidley).

Size comparison of the giant Pleistocene dormouse *Leithia* alongside *Muscardinus*, the common dormouse.

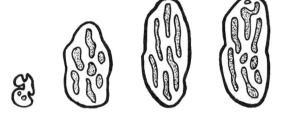

A series of teeth from four mylagaulids, showing the increase in size and complexity of P$_4$ from the early to the late Miocene. (After Shotwell).

lives in a burrow need a pair of horns? As they do not appear to be present on all individuals, they may have been restricted to males. The teeth of mylagaulids also became highly modified; reduced in number, increased in size and height and in complexity of the enamel pattern. Mylagaulids died out with the disappearance of the forests and the spread of the prairies in the late Miocene.

Sciuromorphs – squirrels, dormice and beavers

The sciuromorphs are basically the many kinds of squirrels (over 250 living species) which inhabit a great variety of habitats; there are tree squirrels, ground squirrels, flying (or more accurately gliding) squirrels and burrowing squirrels (or marmots). They are ancient groups going back to the late Oligocene and since that time have spread to all continents except Australia and Antarctica. Their dentition is low crowned and quite simple by rodent standards; indeed it is not very different from that of the primitive paramyids. However, in squirrels the masseter jaw muscle is much bigger and they can cope with tougher and more varied food that could paramyids.

The glirids (dormice) appeared in the Eocene, possibly from the sciuravid ancestry, and they persist to the present in Eurasia and Africa as small arboreal rodents. During the Pleistocene, giant dormice evolved on the islands of Malta and Sicily in the Mediterranean. *Leithia* from Malta was the size of a large rat.

Opposite: restoration of *Epigaulus*; this mylagaulid was about the size of a marmot, had very long compressed claws and a pair of horns on the snout.

Restoration of *Leithia*, a giant dormouse from the Pleistocene of Malta. It is common for island species to be either dwarfs or giants; Malta also has a pigmy elephant.

The castorids (beavers) have a good fossil record in Holarctic regions as far back as the Early Oligocene. They show a progressive increase in the height of the crowns of the teeth (hypsodonty) and many were much more terrestrial than the living aquatic beavers of North America and Europe. *Steneofiber* was abundant in the early Miocene of central France, the best material coming from the famous lake-deposited limestones of St Gerand-le-Puy. Closely related to the European *Steneofiber* was the American *Palaeocastor*, found in abundance in the semi-arid upland sandy Miocene deposits of Nebraska. This beaver was responsible for the devils corkscrews (*Daimonelix*). These extraordinary burrows comprised a large living chamber about 2·5 metres below the ground surface at the end of a tight corkscrew. The walls of the corkscrew have preserved on them the tooth and claw marks of the beaver, and occasionally remains of the animals are found in the burrows.

During the Pleistocene giant beavers evolved in both North America and Europe. *Castoroides* ranged from Alaska to Florida, and was especially abundant around the Great Lakes. It lived in lakes and ponds bordered by swamps and is unlikely to have felled trees; it had short legs and large webbed feet, suggesting it was a powerful swimmer. Adults were as large as a black bear, probably reaching 200 kg in weight and a length of 2·5 m. In spite of its great size *Castoroides* had a brain scarcely larger than that of a living beaver. The contemporary European equivalent was *Trogontherium*, though in size it was only slightly larger than the living beaver; however, it appears to have had a split upper lip like a manatee and this may have been an adaptation to feeding on water plants.

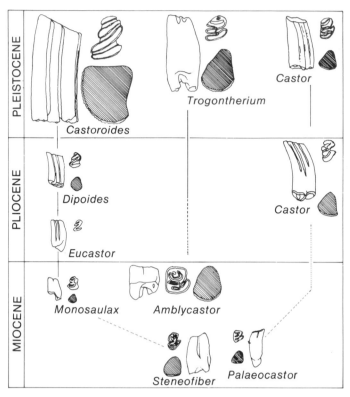

Evolution of cheek teeth in beavers, showing the great increase in hypsodonty or height of the teeth, an adaptation to gnawing woody fibres. Alongside each a transverse section through the incisor tooth. (After Stirton).

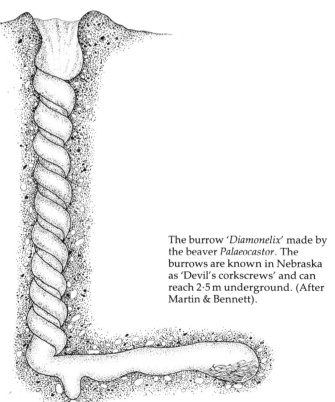

The burrow '*Diamonelix*' made by the beaver *Palaeocastor*. The burrows are known in Nebraska as 'Devil's corkscrews' and can reach 2·5 m underground. (After Martin & Bennett).

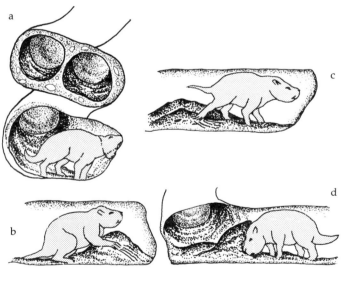

Reconstruction of the mode of excavation of the corkscrew burrow '*Diamonelix*'.
a. spiral burrow is excavated using the incisor teeth;
b. and c. the loose earth is removed with the front and hind feet, or as in (d) is pushed back with the head. (After Martin & Bennett).

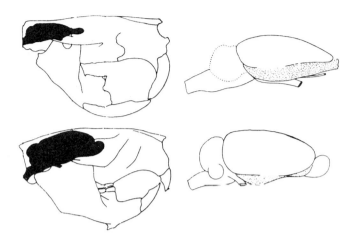

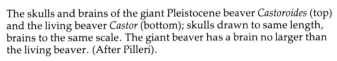

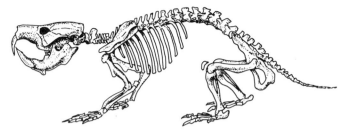

Skeleton of the Miocene beaver *Palaeocastor* from Nebraska. Head and body about 25 cm long. (After Peterson).

The skulls and brains of the giant Pleistocene beaver *Castoroides* (top) and the living beaver *Castor* (bottom); skulls drawn to same length, brains to the same scale. The giant beaver has a brain no larger than the living beaver. (After Pilleri).

Restoration of *Palaeocastor*, a small Miocene beaver from Nebraska; it was about the size of a muskrat and made deep corkscrew burrows.

Theridomorphs – extinct European rodents

This group brings together forms that have hystricomorph jaw muscles and a sciurognathous jaw. The pseudosciurids are a small family known from the European Eocene which in appearance and perhaps in origin form a link between the primitive paramyids and the distinct theridomyids.

Theridomyids were the dominant rodents of the early Tertiary in Europe; small scampering mammals living as do mice and voles today. Their grinding teeth had usually five transverse ridges of enamel enabling them to cope with a wide variety of tough vegetation. In some Oligocene forms such as *Issiodoromys* the number of ridges has been reduced to two and the teeth were very high crowned; the roots remained open in some and the teeth continued growing throughout life. Another feature of *Issiodoromys* was its very large inflated auditory bulla, a characteristic associated with rodents that live today in deserts and steppe terrain. Hearing is their most important sense and the high hypsodont teeth enable them to cope with hard, dry vegetation. All these theridomyids are extinct, but the living anomalurids or scaly-tailed gliding 'squirrels' of Africa may be their descendants.

Myomorphs – hamsters, mice, rats and voles

There are over 1000 species of these rodents living today, that is over a quarter of all living mammal species. An immensely successful and diverse group, but one which only became ubiquitous within the past five million years. The distinctive characteristic of myomorphs is the presence of a myomorphous masseter musculature combined with a sciurognathous jaw. The number of cheek teeth is progressively reduced, the premolars are frequently lost and the molars reduced to three in each row, very occasionally to two or even one.

Paracricetodon from the early Oligocene deposits of Europe is one of the earliest myomorphs and probably descended from a sciuravid. In these early mymorphs cusps are the dominant feature of the molar teeth, linked by slender transverse ridges of enamel. The Oligocene and Miocene record of myomorphs is not spectacular; a sprinkling throughout of families and genera on most continents. However, towards the end of Miocene times things began to change dramatically, culminating in three great explosive waves of speciation. Firstly the cricetids (hamsters and New World mice); they retained cuspidate teeth and diversified geographically (reaching Madagascar, Africa and South America), altitudinally (reaching 5500 metres in Peru) and ecologically from the arctic to the tropics. The cricetines or hamsters are short-tailed, short-faced burrowers and are far outnumbered by the hesperomyines or New World mice which today exceed 350 species – a diversification that appears to have occurred since the early Pleistocene less than two million years ago.

Example of gradual evolutionary change in a series of vole teeth from *Mimomys occitanus* in the Pliocene through two genera and eight species to the living *Arvicola terrestris*. Over this period of about four million years the teeth have gradually become more highly crowned and the triangularity of the cusps more prominent. (After Chaline).

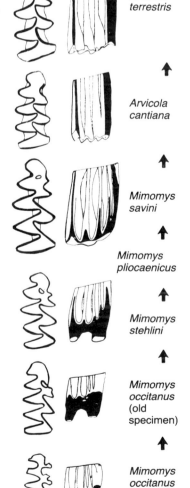

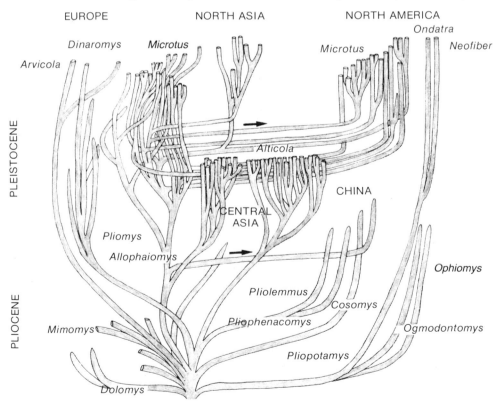

Phylogenetic tree of voles, showing the complexity of their evolutionary diversity during the Plio-Pleistocene. (After Chaline).

The second great speciation wave gave rise to the microtids or voles. In these the dentition becomes progressively more hypsodont and the teeth prismatic, that is with parallel sides and continuously open roots; the enamel pattern on the cheek teeth acquires a characteristic zigzag pattern. The microtids probably originated from cricetids in the late Pliocene from a form close to *Mimomys*. From this evolved in latest Pliocene *Allophaiomys*, the ancestor of the *Microtus* (the field and common vole) lineages. Using the changes in the zigzag

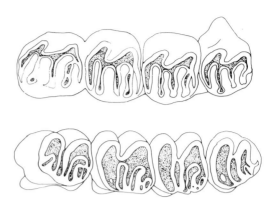

Upper cheek dentition of *Theridomys* (top) and of *Issiodoromys* (bottom), both from the Oligocene of Europe. *Theridomys* teeth have greatly reduced ridges but the crowns are high. (After Stehlin & Schaub).

Upper cheek teeth (M^{1-3} : 8 mm) of *Paracricetodon* from the early Oligocene of France; this is one of the earliest known myomorph rodents with the four cusp pattern still dominant over the slender transverse ridges. (After Stehlin & Schaub).

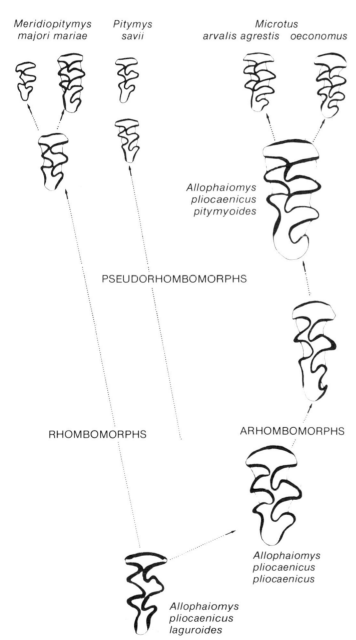

Meridiopitymys majori mariae *Pitymys savii* *Microtus arvalis agrestis oeconomus*

Allophaiomys pliocaenicus pitymyoides

PSEUDORHOMBOMORPHS

RHOMBOMORPHS ARHOMBOMORPHS

Allophaiomys pliocaenicus pliocaenicus

Allophaiomys pliocaenicus laguroides

Evolution of the third molar tooth in voles from a late Pliocene *Allophaiomys* through to a diversity of recent forms. (After Chaline).

Restoration of the vole *Allophaiomys*, the Pliocene ancestor of the *Microtus* lineage. *Allophaiomys* is known in early and mid Pleistocene deposits from Italy through Romania, to the Ukraine.

pattern of the first lower and last upper molar, the diversification of voles can be traced in the Holarctic region throughout the Pleistocene. Over 100 vole species are living today and testify to their phenomenally successful adaptation to grazing in the temperate zones of the northern hemisphere. The largest microtine is the muskrat *Ondatra*, a sub-aquatic American genus economically.important for its fur, known as musquash.

The third great radiation of myomorphs was that of the murids – the rats and mice. Murids probably originated in south-east Asia from late Miocene cricetids. Their fossil record is poor and entirely Old World; they are not particularly common until post-Pleistocene times when they very rapidly became worldwide in distribution, unwittingly aided by mankind. Murids have pointed snouts and long scaly tails; their teeth are usually cuspidate and rooted and their diet is essentially vegetarian but often becomes omnivorous and occasionally insectivorous. The harvest mouse (*Micromys minutus*) is not only the smallest rodent but also one of the smallest mammals (weight about 5 g). Two genera are closely associated with man, *Mus* (mouse) and *Rattus* (rat). Both are recorded from the mid Pleistocene hominid site of Choukoutien near Peking in China. Since that time they have travelled everywhere with man and lived off his food supplies, particularly his grain. Today there are probably more than 500 species of rat-sized murids, but the one to cause man most trouble is the black rat, *Rattus rattus*; it is the carrier of the plague or black death, which is due to a bacillus that breeds in the gut of the rat flea. Rats have been responsible for an enormous toll of human life; in the Middle Ages in Europe the plague is estimated to have killed 25 million people – more than a quarter of the total population. In the present age rats still despoil vast quantities of grain around the world. As species, the success of *Rattus rattus* and its relative *Rattus norvegicus*, the brown rat, can hardly be less than that of *Homo sapiens*.

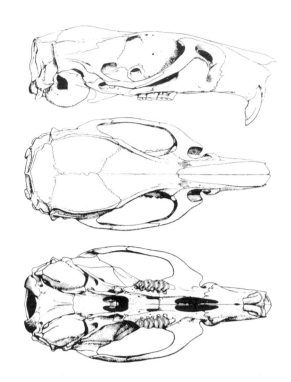

Skull of *Rattus norvegicus*, the brown rat; seen in lateral, dorsal and ventral views. (After Ellerman).

Phiomorphs – Old World porcupines and their relatives

This is a grouping to bring together hystricognathous and hystricomorphous rodents of the Old World. The earliest and most primitive are the phiomyids from the Oligocene of Egypt and most of the descendants are African with a few Eurasian representatives. Phiomorphs share a number of distinctive features, some probably primitive and others probably derived. The number of cheek teeth is usually four on each side of each jaw, P_4 and M_{1-3}, and two of the ear ossicles (malleus and incus) are fused, as they are in most hystricognathous rodents. The phiomyids from the early Oligocene of the Fayum Depression in Egypt are known on the basis of many dentitions, and although there are no skulls they clearly represent a diverse rodent fauna. The dentitions are low to medium-crowned with between three and five transverse ridges. Later phiomyids are known from the Miocene of Kenya where they are accompanied by thryonomyids (cane rats) and bathyergids (mole rats), two extant phiomorph stocks. Africa may have been the centre of radiation for phiomorphs, though as our knowledge of Asiatic faunas increases, these are rapidly becoming strong contenders. The other phiomorph family to consider is

Cheek teeth of phiomorph rodents. Top: $P^4 - M^2$ (9 mm) of *Phiomys* from the Oligocene of the Fayum, Egypt. Middle: $M_1 - M_2$ (11 mm) of *Sivacanthion*, a porcupine from the mid Miocene of Pakistan, with very complex enamel pattern on the low crowned teeth. Bottom: $P_4 - M_3$ (35 mm) of *Hystrix*, an Old World porcupine. (After Stehlin & Schaub).

the Hystricidae, the Old World porcupines. The fossil record of this family extends through southern Asia, the Mediterranean and Africa with the oldest known species being *Sivacanthion* from the mid Miocene of Pakistan. It

Restoration of *Sivacanthion* from the mid Miocene of Pakistan. Although the oldest known porcupine, it is probably on an evolutionary sideline and not a direct ancestor of living porcupines.

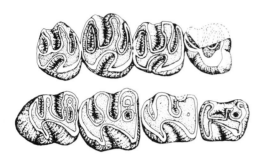

Top, $P^4 - M^3$ (50 mm) and bottom, $P_4 - M_3$ (60 mm) of *Platypittamys* from the early Oligocene of Patagonia, South America. The teeth show the primitive caviomorph pattern. (After Wood).

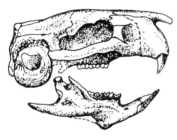

Skull of *Platypittamys* (length 4 cm). Note the small infra-orbital foramen anterior to the eye, with space only for a nerve and not a strand of the masseter muscle. (After Wood).

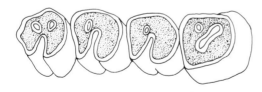

$P^4 - M^3$ (1 cm) of *Cephalomys*, an early Oligocene cavioid from Argentina with very primitive and simple enamel pattern on the teeth. (After Stehlin & Schaub).

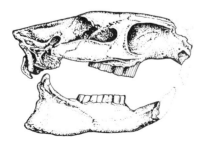

Skull of *Telicomys* (length 30 cm) from the Pliocene of Argentina. (After Kraglievich).

is a member of the only phiomorph family with a good non-African record.

Caviomorphs – New World porcupines and their relatives

This grouping accounts for all the hystricognathous and hystricomorphous rodents of the New World. Caviomorphs and phiomorphs do not share any derived characters in common and their relationship is hotly debated; it will be discussed in the terms of the biogeography of South America in Chapter 11. The caviomorphs are a large and very diverse group, with over 150 genera, fossil and living, arranged in at least a dozen families, but all falling fairly clearly into four superfamilies. They are found throughout South America; about half the families have or have had representatives in the West Indies and Central America; only the porcupines have invaded North America. Caviomorphs usually have high crowned teeth with well-developed transverse ridges of enamel on the cheek teeth. They vary greatly in habitat, form and size, and include the largest known rodents.

The early Tertiary deposits in South America contain no rodents; they make their debut in the far south of Patagonia along with native mammals in the early Oligocene Deseadan fauna. The rodents in these beds all possess many features in common, yet among them can be recognized the beginnings of the four superfamilies that were to evolve on the continent over the ensuing 35 million years.

Platypittamys (an octodontoid) is the most primitive of these caviomorphs, with a small infra-orbital foramen, too small to accommodate a strand of the masseter muscle. Contemporary with it was the cavioid *Cephalomys* in which the hystricomorph masseter condition was fully developed. The major stocks of caviomorphs are the octodontoids (mostly rat-like in appearance); the cavioids with the guinea pigs, pacas and the largest living rodent, the capybara; the chinchilloids or chinchillas and the erethizontoids or porcupines. Throughout the later Tertiary these stocks have radiated and evolved into many species in South America – a good example of the way in which rodents rapidly exploit an area and diversify into innumerable niches. The cavioids produced the most diverse stocks; basically they comprise the guinea pigs, but also include the capybara; its latin name is *Hydrochoerus*, meaning 'water pig', which befits a pig-like animal that lives close to swamps and lakes. Reaching about 50 kg in weight, the capybara is, however, small compared with some of its fossil relatives. The Pliocene genus *Protohydrochoerus* was about the size of a tapir. However, the real giants among these rodents were extinct members of the family Dinomyidae; *Telicomys* in the late Miocene and Pliocene reached the size of a small rhinoceros.

The erethizontoids are the New World porcupines; they have few resemblances to the Old World porcupines other than the possession of quills, and even these differ in structure. Th erethizontids are arboreal and since their migration into Central America in the Pliocene they have radiated across the forested areas of North America as far as Alaska.

Restoration of *Protohydrochoerus*,
the giant capybara from the
Pliocene of South America.

Scale drawings of large fossil rodents and living mammals. On the
right, the Pliocene capybara *Protohydrochoerus* and beside it a tapir. On
the left, the gigantic Pleistocene *Telicomys* which stood as high as a
small rhinoceros. *Telicomys* may have been more closely related to the
false paca *Dinomys* than to the capybaras, although in proportions and
build it may have resembled capybaras.

Lower cheek dentition (P$_4$ – M$_3$: 3 cm) of *Erethizon*, the North American porcupine. (After Stehlin & Schaub).

Lower dentition (P$_4$ – M$_3$: 11·5 cm) of *Protohydrochoerus* from the Pliocene of Argentina. The teeth form a continual row of transverse enamel ridges for shredding vegetation. (After Stehlin & Schaub).

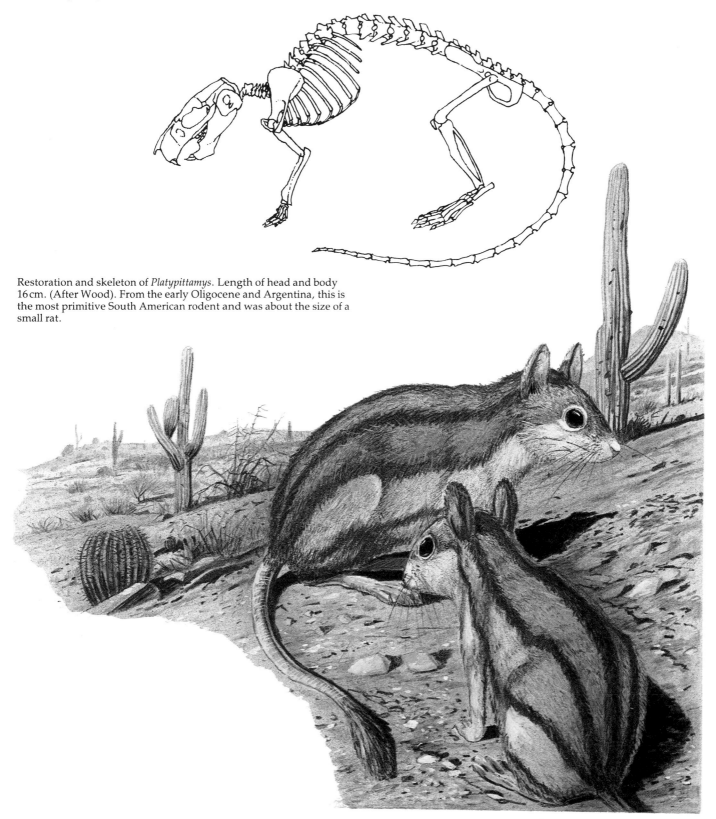

Restoration and skeleton of *Platypittamys*. Length of head and body 16 cm. (After Wood). From the early Oligocene and Argentina, this is the most primitive South American rodent and was about the size of a small rat.

Lagomorphs – pikas, rabbits and hares

The pikas, rabbits and hares form a small order of about a dozen living and some 50 fossil genera with a record going back to the late Palaeocene. They resemble rodents in some ways – they are all medium to small-sized gnawing mammals; they have persistently growing incisor teeth and cheek teeth with transverse ridges. But there the superficial similarities of rabbits and rodents end; they do not appear to be in any way related. Lagomorphs have two pairs of incisor teeth in the upper jaw (only one pair in rodents) and the enamel completely surrounds the tooth (restricted to the front of the teeth in rodents). The cheek teeth usually number five or six (always five or less in rodents) and there are only two transverse ridges on each tooth (usually more in rodents). They are widespread vegetarian feeders on grasses in the plains and on shrubs in rocky tundra and desert terrains.

The earliest fossil lagomorphs occur in Mongolia and China with small mammals such as *Eurymylus*; already specialized as a lagomorph, it could have evolved from either an insectivore or an ungulate. The two living families are the ochotonids (pikas) and leporids (rabbits and hares). The earliest ochotonids are found in mid Oligocene rocks in eastern Asia and the family has a reasonable fossil record in Eurasia and North America through the later Tertiary, with an excursion into Africa in the Miocene. Recent work on the blood of pikas suggests that their closest relatives (after rabbits and hares) are tree shrews and carnivores; but the

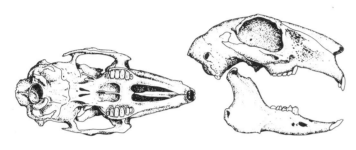

Skull of the brown hare *Lepus europaeus* (length 9.5 cm).

Cheek dentition of *Eurymylus*, the earliest known lagomorph, from the late Palaeocene of Mongolia; although worn, the typical lagomorph transverse enamel folds are discernable. (After Wood).

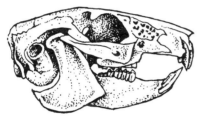

Skull of lagomorph *Palaeolagus* (length 5 cm), an early rabbit from the Oligocene of western USA. (After Wood).

Cheek dentition of *Palaeolagus* with simple paired transverse enamel folds on each tooth, a characteristic feature of leporids. (After Wood).

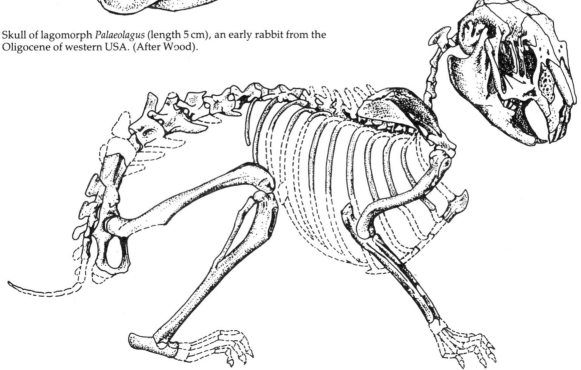

Skeleton of *Palaeolagus*. The limb proportions suggest it may have been less adapted to leaping than later leporids. (After Wood).

experimenters did not test the proximity of rodents.

The leporids can be traced back to the late Eocene in both Asia and North America and have been abundant and varied in both the Old and New Worlds through most of the Tertiary; with the introduction by man of rabbits into Australasia they have become virtually worldwide. The fossil remains are usually jaw fragments and teeth, but occasionally nearly complete skeletons turn up, as with *Palaeolagus*, an early rabbit from the Oligocene of western USA. In the Tertiary evolution of leporids we can trace a simplification of tooth patterns, an increase in the height of the crowns of the cheek teeth and the development of longer hind legs, enabling them to achieve a leaping gait.

Medieval French drawings of rabbits being hunted in a rabbit garden by ladies; originally published in 1393. (After Nachtsheim).

Restoration of *Palaeolagus*, an Oligocene rabbit from western USA.

CHAPTER 10
Early rooters and browsers

The mammals discussed here are those that formed the first great wave of vegetarians. They ranged from rabbit- to elephant-sized creatures; most were hoofed but some retained claws. Of the ten stocks only two survive to the present – the conies and the elephants. All the others thrived in the Palaeocene and Eocene with a few persisting into the Oligocene. They are presently known very largely from the northern hemisphere; only embrithopods and the two living groups have African domiciles. They appear to have occupied a wide variety of habitats; a few were arboreal and several were semi-aquatic. Some were probably omnivorous feeders, others may have had specialized diets, but the great majority were rooters, feeding on tubers. Clawed feet and large canines were used to uproot the tubers and the generally broad low crowned cheek teeth used to mash the vegetation. Thus we see that 'herbivore' and 'ungulate' (literally 'hoofed') are not synonymous terms; some specialized herbivores have claws and some specialized relatives of the carnivores are herbivorous.

Three of the orders (Arctocyonia, Acreodi and Condylarthra) contain the ancestors of most of the herbivorous mammals. The elephants are for convenience dealt with here, the South American herbivores and the odd and even toed ungulates in the following two chapters.

EARLY ROOTERS AND BROWSERS

Order	
Taeniodonta	*Stylinodon*
Pantodonta	*Titanoides*
	Coryphodon
	Barylambda
Arctocyonia	*Protungulatum*
	Kopidodon
Acreodi	*Ichthyolestes*
	Harpagolestes
	Mesonyx
	Andrewsarchus
Tillodontia	*Trogosus*
Dinocerata	*Uintatherium*
Embrithopoda	*Arsinoitherium*
Condylarthra	*Hyopsodus*
	Phenacodus
	Tetraclaenodon
	Meniscotherium
Hyracoidea	*Titanohyrax*
	Heterohyrax
Proboscidea	
Moeritheroidea	*Pilgrimella*
	Moeritherium
Barytheroidea	*Barytherium*
Deinotheroidea	*Deinotherium*
Elephantoidea	
Mastodonts	*Palaeomastodon*
	Phiomia
	Mammut
	Gomphotherium
	Platybelodon
	Anancus
	Cuvieronius
	Stegodon
Elephants and	
mammoths	*Primelephas*
	Loxodonta
	Elephas
	Mammuthus

Taeniodonta

Taeniodonts appear to have originated as hedgehog-like animals from insectivorous ancestors close to *Cimolestes* (see p.53); they show a series of trends through their 20-million-year history in Holarctica ending with *Stylinodon* in the late Eocene. Although no complete skeleton is known, it was probably pig-like. We know the limbs were short and stocky and the feet retained claws that could have been used to uproot tubers. The jaws were very powerful; the canines large and rootless and the cheek teeth high crowned (hypsodont) with only a band of enamel (hence the name taeniodont which means banded tooth). Living on subterranean tubers, taeniodonts may have been able to live as aardvarks do today in waterless shrublands and this could explain their relative rarity in local faunas.

Pantodonta

Pantodonts share with taeniodonts an insectivorous ancestry. They are fairly abundant and varied in the early Tertiary sequences of the northern hemisphere. *Titanoides* was probably terrestrial with a bear-like appearance; it had very large canine tusks and its front limbs were mechanically adapted for digging and equipped with claws. The cheek teeth are typically pantodont i.e. broad crowned with cross ridges (lophs). *Coryphodon* was tapir-like in appearance; the short

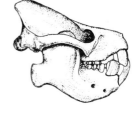

Left: skull of *Stylinodon* (length 27 cm), a taeniodont from the late Eocene of North America; note the very deep mandible and the enlarged rootless chisel-edged $\underline{C}$, $\overline{C}$ and P_1 tusks. (After Patterson). Right: stylinodont front (left) and hind (right) foot; the larger front foot has strong claws. (After Matthew and Gazin).

Scene in the mid Palaeocene of the Rockies with a wombat-like
taeniodont *Ectoganus* (foreground) and a pantodont *Pantolambda*.

stocky limbs had unfused radius/ulna and tibia/fibula,
which enabled the feet to be turned sideways and the
five-toed feet were hoofed. The skull of *Coryphodon* was
large with, in males, very big canine tusks. The brain,
however, was very small, only 90 g in an animal of

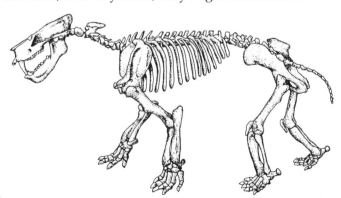

Skeleton of *Titanoides* (length 1·6 m) from 55-million-year-old deposits
in Montana. (After Simons).

Restoration of *Titanoides*, a small bear-like pantodont from the late
Palaeocene of North America.

around 500 kg; this is one of the smallest brain/body weight ratios known among mammals. *Barylambda* was hippo-like, standing one metre at the shoulder. The pantodonts can thus be seen as essentially large, bulky, semi-aquatic feeders on roots, tubers and aquatic vegetation.

Pantodont skulls. Right, top: early Eocene *Coryphodon* (length 56 cm; middle: the late Palaeocene *Barylambda* (length 48 cm); bottom: the late Palaeocene *Titanoides* (length 43 cm). Three dietary specializations are suggested by the elongate canine in *Titanoides*, the short canine in *Barylambda* and the enlarged shearing canines in *Coryphodon*.

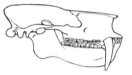

Restoration of *Coryphodon*, a tapir-like pantodont from the early Eocene of North America and Eurasia.

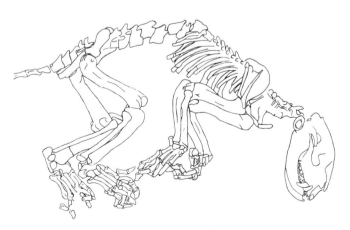

Skeleton of *Kopidodon* (length 60 cm), 45 million years old. (After von Koenigswald).

Arctocyonia

Protungulatum from the late Cretaceous of Montana is the earliest known eutherian mammal to show herbivorous tendencies; the cheek teeth broaden transversely and the upper molars have a well-developed internal cingulum. These features suggest an animal adapted to crushing vegetation rather than slicing meat. Arctocyonids of the Palaeocene varied from sheep- to bear-sized creatures; they had small brains and slender limbs with clawed feet. Their dentition shows a tendency for the cheek teeth to square up with the development of a hypocone. Of the 30 or so genera in the family, most were probably omnivores. Closely allied to the arctocyonids, if not actually a member of the family, is *Kopidodon* from the mid Eocene oil shales of Messel in Germany. Several near complete skeletons have been found of this beast which was about the size of a racoon, and used its well-clawed feet for climbing through the trees where it lived. The teeth of *Kopidodon* have a very unusual wear pattern which appears to have been produced by eating a special type of fruit. From the arctocyonids probably evolved most, if not all, the other condylarths, and hence through them, most of the herbivores.

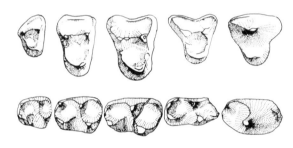

Upper and lower dentition of *Kopidodon* (length P_3–M_3 = 27 mm). The wear pattern suggests a diet of fruit. (After von Koenigswald).

Restoration of the racoon-like arctocyonid *Kopidodon* from the mid Eocene oil shales of Messel in Germany.

Reconstruction of the skull of *Protungulatum* (length 6 cm), the most primitive known ungulate from the late Cretaceous of the Rockies. (After Szalay).

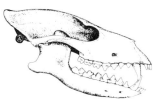

Lateral view of the mandible of *Protungulatum*; P_2–M_3 = 21 mm). (After Kielan-Jaworowska, Bown & Lillegraven).

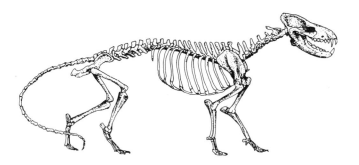

Skeleton of *Mesonyx* (length 1·5 m), a mid Eocene mesonychid from western USA. (After Scott).

Acreodi

The only members of this order are the mesonychids that were descended from triisodont arctocyonids which had broad bunodont cheek teeth; they vary from fox-sized to gigantic carnivore-like beasts and are known from mid Palaeocene times, some persisting into the Oligocene. *Mesonyx* was superficially wolf-like; it had triangular upper molars and shearing talonids on the lower molars. The feet were five-toed but with small hoofs rather than claws. *Harpagolestes* was an omnivore with a massive skull, strong jaws, very large canine tusks and broad cheek teeth; like many mesonychids it was probably a scavenger and perhaps able to crush bones for marrow as do hyaenas. The gigantic and formidable *Andrewsarchus* from the early Oligocene of Mongolia had a skull which was over one metre in length; it must have been a terrifying beast as it scavenged for food.

The detailed anatomy of the base of the skull and some of the dental features in mesonychids have similarities with those of early cetaceans and it seems likely that it was from this stock that the whales evolved. *Ichthyolestes* from the mid Eocene of Pakistan is sometimes classified as a cetacean and at other times as a mesonychid, but was originally described as a fish-eating creodont.

Restoration of *Mesonyx*, a 50-million-year-old mesonychid with the appearance of a large wolverine.

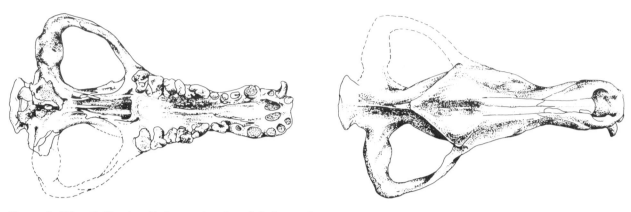

Above, skull (length 83 cm) and below, restoration of *Andrewsarchus* a
gigantic omnivorous scavenger from the late Eocene of Mongolia.
(After Osborn).

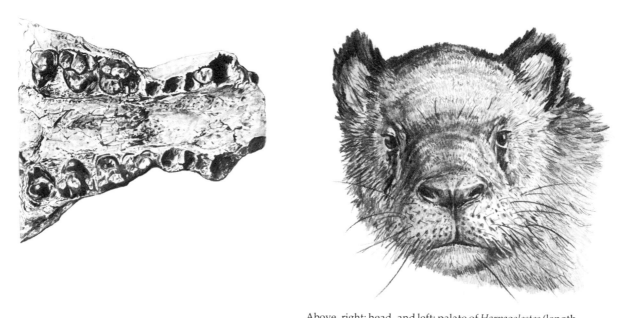

Above, right: head, and left: palate of *Harpagolestes* (length
I–M^3 = 13.5 cm), a mesonychid from the late Eocene of Mongolia. This
large bear-sized omnivore had massive canine teeth (sockets only
preserved) and a heavy crushing molars. (After Szalay and Gould).

Tillodontia

Tillodonts are known from about ten genera in the Palaeocene and Eocene of North America and eastern Asia, and a couple of genera from Europe. No skeletons have been found and we still know nothing of their relationships or mode of life. The simple three-cusped upper molars with a broad inner shelf suggests arctocyonid affinities. The very enlarged and rootless second pair of incisors is reminiscent of moeritheres (see p.143) and even rodents. *Trogosus* was about the size of a brown bear, had bear-like limbs with plantigrade five-toed clawed feet.

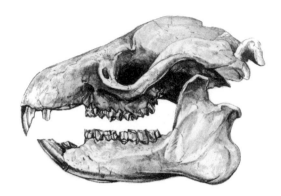

Skull of tillodont *Trogosus* (length 30 cm) from the mid Eocene of western USA; a bear-sized skull with very rodent-like incisor dentition. (After Gazin).

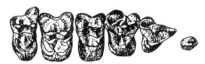

Upper P^2–M^3 (= 7 cm) of *Esthonyx*, an early Eocene tillodont from Wyoming. Note the molarization of P^4. (After Gazin).

Restoration of *Trogosus*, a large bear-like tillodont with clawed feet.

Dinocerata

About a dozen genera of these 'terrifying horns' are known from North America and eastern Asia; most were large and some very large. *Uintatherium* (named after the Uintah tribe of Indians in Utah) was as large as an African rhinoceros and of similar build. The skull was adorned with three pairs of bony protuberances; the anterior pair on top of the nose may have been sheathed in dermal horn like a rhinoceros. The posterior pair were large, up to 25 cm high and probably, like the middle pair, not sheathed in horn. To add further to their striking appearance, especially in the males, uintatheres had a pair of 15-cm-long sabre-like upper canines. The upper incisors were lacking and the lower ones very small; this together with the shape of the anterior part of the mandible suggests the tongue may have been specialized for food gathering, perhaps browsing on some particular type of vegetation. The cheek teeth have V-shaped crests and could have been used for mashing food in a similar way to those of rhinoceroses. Asiatic genera lack both protuberances and sabre canines.

In attempting to reconstruct the mode of life of these great beasts, we are without a close modern parallel. Rhinoceroses, giraffes, deer and cattle have skull protuberances but lack sabre canines. However, some antlerless deer, in particular the males, do have very large upper canines; these are seen in *Muntiacus* (muntjac), *Moschus* (musk deer) and *Hydropotes* (Chinese water deer). These deer are all very small and use their canines in dominance contests and in slashing edible bark off trees. Dinocerates probably had their origin in arctocyonids and their ecological niches as massive herbivores were taken over in the Oligocene by brontotheres.

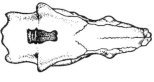
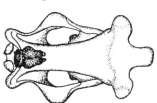
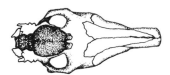

Skulls with brain outlined. *Uintatherium* top, *Brontotherium* middle and *Equus* (horse) bottom. All drawn to the same skull length, which emphasizes the minute size of the brain of *Uintatherium* compared with that of the horse. (After Marsh).

Left and right upper dentition of *Uintatherium* (P^2–M^3 = 15 cm) with characteristic V-shaped crests. (After Marsh).

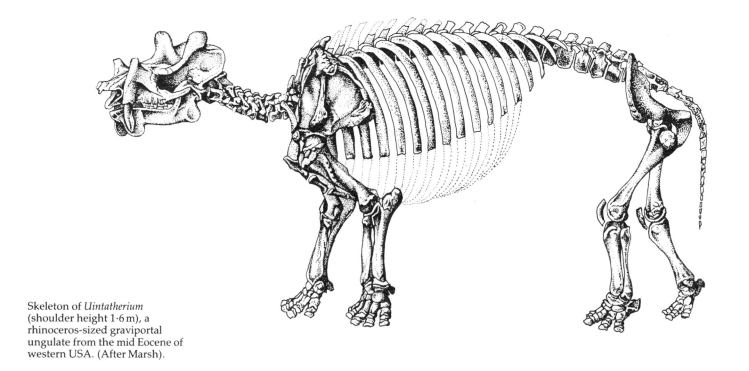

Skeleton of *Uintatherium* (shoulder height 1·6 m), a rhinoceros-sized graviportal ungulate from the mid Eocene of western USA. (After Marsh).

Restoration of *Uintatherium*, a 50-million-year-old horned ungulate, ecologically comparable with the living African rhinoceros.

Embrithopoda

In early Oligocene times in Egypt lived the magnificent horned *Arsinoitherium* (named after Queen Arsinoe, wife of King Ptolemy, who had a palace near the site where the fossil was found). *Arsinoitherium* was the African counterpart of the dinocerates; it was a large rhinoceros-sized creature, heavily built with graviportal limbs and five-toed hoofed feet. A pair of massive but hollow 'horns' arises from most of the skull roof; these are fused at the base, rounded in young individuals and females, but larger and pointed in male adults. Some have traces of the course of blood vessels indicating that for some of the time at least they were covered with skin, perhaps in

a similar fashion to the ossicones in the giraffe. *Arsinoitherium* had additionally two small 'horns' immediately behind the large pair. The brain was very small, simple and rather similar to that of early sirenians. The dentition is complete: all 44 teeth are present and form a continuous arcade; this is unusual in herbivorous mammals, which usually have a gap (the diastema) separating the anterior cropping teeth and the posterior masticatory teeth. In *Arsinoitherium* the cheek teeth are unusually highly crowned for an Oligocene mammal, suggesting perhaps the ability to cope with very tough vegetation.

There is a second but poorly known species in the

Restoration of *Arsinoitherium*, a 35-million-year-old rhinoceros-like ungulate from Egypt.

same Egyptian deposits, and it was as large as an
elephant. Recently arsinoitheres have been found in
Libya, Romania and Turkey, with a possible ancestor in
Mongolia. As yet the mode of life of arsinoitheres
remains unknown. Though rare and the bones isolated,
a nearly complete skeleton has been assembled. Skulls
and mandibles of 47 individuals are recorded from the
fluvial sediments of Fayum in Egypt; of these 18 are
juvenile and seven young adult – a high proportion of
immature animals. Did it live among the dense gallery
forests which bordered the river channels, or was its real
home beyond on the steppe, feeding on tougher
vegetation?

Upper dentition cheek of *Palaeoamasia*, an embrithopod from the
Oligocene of Turkey. The cheek teeth (P^3–M^3 = 12 cm) are bilophodont
and rather like rhinoceros teeth. (After Sen and Heintz).

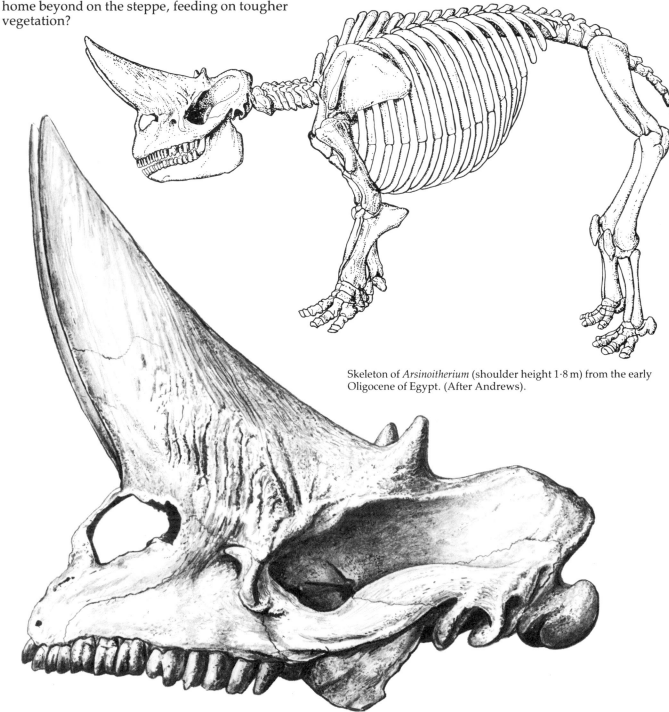

Skeleton of *Arsinoitherium* (shoulder height 1·8 m) from the early
Oligocene of Egypt. (After Andrews).

Skull of *Arsinoitherium* (length 75 cm); an adult male with long pointed
horns showing traces of blood vessels near the horn base. (After
Andrews).

Condylarthra

This order comprises a loose grouping of five or six families of early herbivores, some ancestral to later orders and others dead ends. They have been found mainly in Palaeocene and Eocene deposits of North America and Europe, but our knowledge of Asiatic forms is increasing rapidly.

Hyopsodontidae comprise about 15 genera of the order. *Hyopsodus* itself was about the size of a hedgehog and rather insectivore-like; the dentition was complete and without a diastema, the canine was reduced and the molars bunodont. The spreading clawed feet suggest a semi-arboreal life style. Hyopsodonts are close to the origin of artiodactyls (see p.208).

Phenacodontidae comprise only about four genera of which *Phenacodus* is the best known and *Tetraclaenodon* from the early Palaeocene is the most generalized and primitive form. A typical phenacodont would be about sheep-sized, with a long tail, short legs in which the feet were free to rotate sideways and each of the feet had five hoofed digits. The brain was small and primitive; the dentition was complete with large canines and molar teeth, each of which had six low cusps. Phenacodonts stand close to the origin of perissodactyls (see p.190) and probably also gave rise to meniscotheres, another small family of condylarths with a selenodont dentition; selenodont molars have crescent-shaped cusps which are common in later more advanced herbivores like deer and cattle.

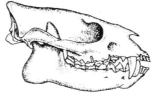
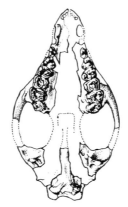

Skull of *Tetraclaenodon* (length 13 cm), a primitive phenacodontid condylarth from the mid Palaeocene of western USA. (After Matthew).

Hind foot of *Tetraclaenodon*; the five-toed hoofed digits are similar to those of the phenacodontid horse ancestors.

Below: skeleton of *Meniscotherium* (length 60 cm) a hyracoid-like condylarth from the early Eocene of North America. (After Gazin).

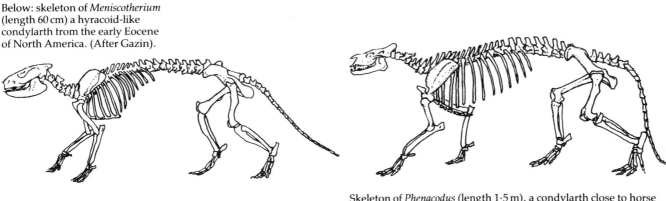

Skeleton of *Phenacodus* (length 1·5 m), a condylarth close to horse ancestry from the late Palaeocene of North America. (After Gazin).

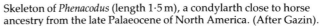

Above: skeleton of *Hyopsodus* (length 30 cm) from the mid Eocene of Wyoming. (After Gazin).

Restoration of *Hyopsodus*, a small condylarth close to artiodactyl ancestry.

Restoration of *Phenacodus*, a sheep-sized condylarth with five hoofed
digits on each foot.

Hyracoidea

'The conies are but a feeble folk, yet make they their house in the rocks' PROVERBS 30, 26.

The conies of the Bible are rabbit-sized animals with many peculiar features and few relatives. They, with the rock and tree hyraxes of Africa are the sole survivors of a group of herbivorous mammals which were diverse and abundant 35 million years ago and appear to have declined slowly since. Unlike rabbits they have short ears, but otherwise resemble them superficially. They have a scent gland in the middle of the back with which they mark territories and they have a terrifying call like that of the banshee or a child in agony. Their skeleton has lots of odd features: nearly 30 vertebrae between head and sacrum where 20 or so is normal in other mammals; their feet are five-toed with a claw on the inner digit of the hind foot and flattened nails on other digits. The mandible often has a deep fossa posterointernally and this is believed to be in some way responsible for their shrill cry. The incisor teeth are enlarged, the lowers used for grooming. The cheek teeth vary with the species from bunodont to selenolophodont and their pattern has caused them to be confused with chalicotheres and allied with South American typotheres.

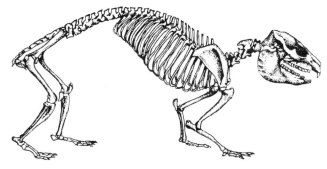

Skeleton of *Heterohyrax* (length 40 cm), the living rock hyrax of Africa. (After Blainville).

The earliest fossil record of hyracoids is of a poorly preserved specimen from the early Eocene of Algeria. Then from the early Oligocene of Egypt we know a rich variety of at least six genera, varying in size from rabbit to tapir (such as *Titanohyrax*) and probably adapted to a wide range of ecological niches. By the Miocene there were fewer kinds, though in late Miocene times they migrated out of Africa into Eurasia (as *Pliohyrax* in Greece and Iran). With the rise of the bovids soon after, they declined further and now have a very limited foothold.

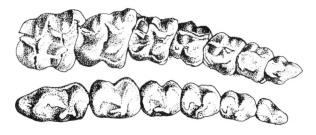

Upper and lower dentition of *Titanohyrax*, a large tapir-sized hyracoid from the Oligocene of Egypt. (After Andrews).

Proboscidea

The elephants of Africa and south-east Asia are the sole survivors of an almost worldwide radiation of gigantic herbivores that started in the early Tertiary, certainly by the Eocene and perhaps during the Palaeocene. The first elephantoid of which we have good evidence is *Moeritherium* from the Eocene of North Africa. Though not considered a direct ancestor of living elephants, it serves well as a morphological archetype; we will examine the differences between the early and living types and discover how the elephant got his trunk.

How the elephant got his trunk – according to Rudyard Kipling.

If there is one overall trend to be detected in elephant evolution, it is increase in weight. *Moeritherium* was already about 225 kg, the size of a pigmy hippo or a tapir. An average African cow elephant weighs 3·6 tonnes (a 16 fold increase); the maximum recorded weight for an African bull elephant standing 4 m at the shoulder is 12 tonnes (a 53 fold increase). The largest known fossil elephant (*Mammuthus trogontherii*) stood 4·5 m and probably weighed about 18 tonnes. As the weight and therefore the volume increased manifold, so there developed problems of locomotion. The legs had to get longer to raise the body clear of the ground; each elephant leg carries a load of 1·5 tonnes or more and if the limb bones were positioned like a dog's hind leg, that is in a S-shape, the energy required to maintain posture would be enormous. For efficiency, the limb is pillar straight, I-shaped, with the weight being taken directly along the bone axis. Elephants never gallop (galloping being a fast gait involving a floating phase when all four feet are off the ground simultaneously). Elephants amble, but with their long legs they can achieve $25 \, \mathrm{km \, h^{-1}}$.

Another problem of weight increase is processing enough food. Fortunately the metabolic rate (i.e. energy consumption per kilogramme unit of body weight) is much less for large than for small mammals. A shrew eats daily about its own weight in food; an elephant eats daily about 1·3 per cent of its body weight, or about 50 kg for a 3·7 tonne animal. The metabolic rate of shrews is about 100 times that of elephants. Moeritheres we may

Restoration of *Mammuthus trogontherii,* the steppe mammoth from the
mid Pleistocene of Europe. This gigantic mammoth stood 4·5 metres at
the shoulder and had tusks 5 metres long.

estimate consumed daily about 1·6 per cent of their body weight, or about 3·5 kg; elephants eat 14 times as much as did moeritheres.

All food has first to be processed in the mouth; mashed and ground by the action of the cheek teeth. In moeritheres there are six cheek teeth in each half of each jaw; these have a crown volume of around 72 cm^3 and assuming a life span of 15 years, this allows for an annual wear rate of 4·8 cm^3. Elephants also have six cheek teeth, but they do not all erupt together as in moeritheres. The first three are small and the fourth begins to function in the fourth or fifth year of life; this is succeeded eight years later by the fifth, and the last molar begins to erupt when the animal is about 25 years old and lasts for another 50 years. By having this cyclic succession, with only one tooth fully functional in each jaw half at a time, the individual teeth can be very large and high crowned. The crown volume of these last three molars is around 4500 cm^3, giving the elephant an annual wear rate of about 65 cm^3, which is again about 14 times that of moeritheres. If elephant teeth were smaller, they would wear out too soon and the animal die of starvation, unable to chew its food.

In the course of evolution, elephants have enlarged their second incisor teeth as tusks; first for food gathering and later for display. A bull elephant's pair of tusks can weigh 200 kg; add to this four large molar teeth, long trunk, big ears, skull and mandible bones, and the head can in total weigh one tonne. With the need to consume 200 litres or more of water daily, this must be acquired with minimum energy output; since the shoulder is 3·5 metres above the ground there are but three possible solutions to the problem. The elephant could get down on its knees and lap water like a carnivore, but in this position it would be highly vulnerable to attack. The elephant could, like the giraffe, carry its head on a long neck, but one tonne of head held three metres from the body would impose an impossible load on the musculo-skeletal system. The third alternative is to keep the head close to the body and develop a tube with which to raise water; and this is the solution adopted by the proboscideans – the proboscis- or trunk-bearing mammals.

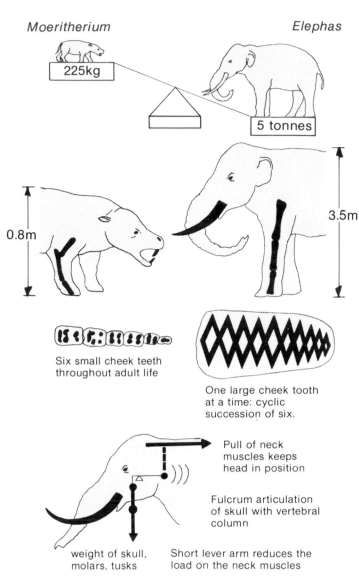

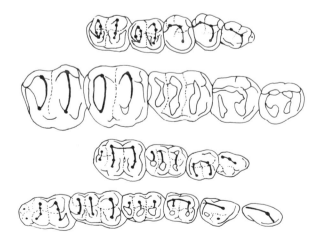

Series of upper dentition of early proboscideans: from the top downward: *Moeritherium* (P^2–M^2); *Deinotherium* P^3–M^3); *Palaeomastodon* (P^3–M^2); *Phiomia* (P^2–M^3). All to same scale. (After Tassy).

Moeritherium and *Elephas* compared and how the elephant really got his trunk. In an African elephant the weight of the skull, with its four molar teeth and long tusks, can exceed one tonne; this weight could not be supported at the end of a long neck, thus the skull must be close to the body; since the legs are long, the mouth is a long way from the ground (and water); hence the need for a trunk to eat and drink.

The earliest proboscideans so far known are grouped in the family Anthracobunidae; all five genera come from the early and middle Eocene of north-east Pakistan and north-west India. *Piligrimella* is typical; it was the size of an African river hog with a weight of about 120 kg. The dentition was complete and without a trace of tusk development. The molars were essentially four cusped and bunodont with incipient cross ridges; the association of the fossils with brackish water deposits suggests they may have fed on soft aquatic vegetation. These unspecialized mammals could include the ancestors of both moeritheres and sirenians.

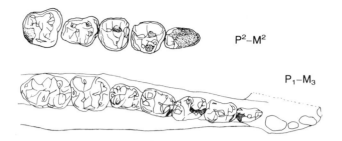

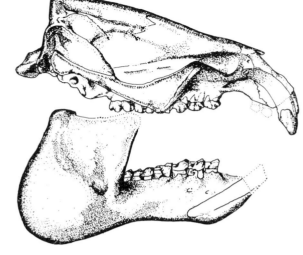

Upper and lower dentition of *Pilgrimella* (P^2–M^2 = 86 mm), the earliest and most primitive known proboscidean from the early Eocene of Pakistan. (After Wells and Gingerich).

Fairly complete skeletons of *Moeritherium* are known from the early Oligocene of Egypt and species have been recovered from Eocene beds in the Sahara from Algeria to Mali. In *Moeritherium* the second upper and lower incisor are clearly incipient tusks. The first premolars are absent and in the mandible both the canine and lateral incisor are also lacking, leaving a diastema behind which is developed the bunolophodont cheek dentition. The body and limb proportions of *Moeritherium* are very similar to those of *Choeropsis* the pigmy hippo, and it would be reasonable to conclude that they had similar modes of life.

Skull of *Moeritherium* (length 44 cm) from the late Eocene of Egypt. (After Osborn).

Restoration of *Moeritherium*, a 40-million-year-old proboscidean from Egypt.

Barytheres and deinotheres

These two minor proboscidean stocks are more closely related to each other than to other proboscideans and they appear to share a common ancestry with the moeritheres.

Barytheres are known only from the Saharan Eocene and were elephantine in size and build. They had two pairs of short incisor tusks in both upper and lower jaw; the upper tusks are vertical the outer pair (I^2) being much larger than the inner pair (I^1); the lower tusks are horizontal, with the inner tusks (I_1) much larger than the outer pair (I_2). This tusk arrangement produced a shearing action not unlike that in the anterior dentition of hippopotamus. A diastema separated the tusks from the battery of bilophodont shearing cheek teeth. *Barytherium* bears many very close resemblances to the Eocene and Oligocene pyrotheres of South America (see p.174).

Deinotheres were similar in build to barytheres; they are known from early Miocene times in Africa and Eurasia, and their last known occurrence is in the early Pleistocene of East Africa. Deinotheres were contemporary with mastodonts (see below) but their distribution is much more restricted; they are known only from East and North Africa, from central and southern Europe, and southern Asia as far east as India. Throughout their 20-million-year history they remained remarkably stable in form and specific differentiation is based mainly on size increase with time. The skull differs in many details from that of elephantoids, but the high nostrils indicate presence of a trunk. There were no

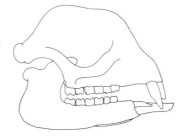

Skull of *Barytherium* (length 85 cm) from the early Oligocene of north Africa. An elephant-sized proboscidean with hippo-like incisors and deinotherium-like molar teeth.

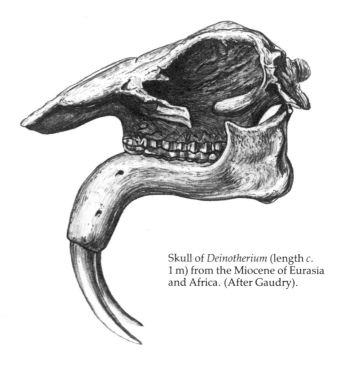

Skull of *Deinotherium* (length *c.* 1 m) from the Miocene of Eurasia and Africa. (After Gaudry).

Upper dentition, P^3–M^3 (length 32 cm) of *Deinotherium*. M^1 has three lophs; P^4, M^1 & M^2 have two lophs each.

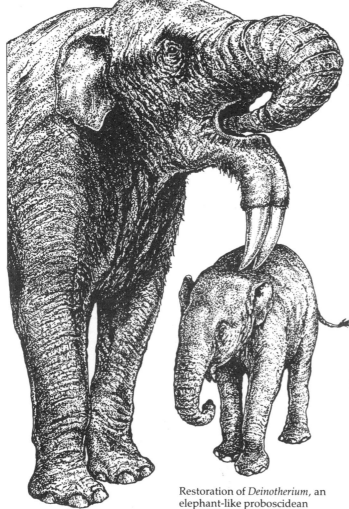

Restoration of *Deinotherium*, an elephant-like proboscidean which lived in Eurasia and Africa from Miocene times through to the Pleistocene.

Upper dentition, P^2–M^3 (length 32 cm) of *Barytherium*. All three molars are bilophodont, but the incisors and premolars differ from those of deinotheres.

Restoration of *Phiomia*, a 35-million-year-old proboscidean from north Africa.

Upper dentition of *Phiomia* (length P²–M³ = 25 cm) from the early Oligocene of Egypt. The molar teeth have four bunodont cusps and talonid; P⁴ is molarized.

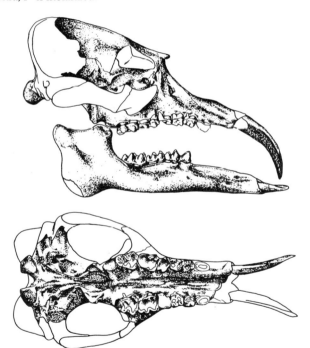

Skull of *Phiomia* (length 52 cm) from the Oligocene of north Africa. (After Osborn).

Upper molar tooth of a bunolophodont mastodont; the cusps are low, blunt and conical with many small accessory cusplets, which, on wear, unite with the main cusps to give a trefoil pattern.

Upper molar tooth of a zygolophodont mastodont; the cusps are transversely keeled, pairs forming lophs, without accessory cusplets.

anterior teeth in the upper jaw, and the single pair of tusks in the mandible was large, downturned and backwardly curved; a few have the tips preserved, and the most likely use would have been for removing the bark from trees. Behind a long diastema the cheek teeth were lophodont, the first molar trilophodont and the others bilophodont. Judging by the tooth wear, the anterior cheek teeth were used for crushing food and the posterior teeth were primarily adapted for shearing; the teeth were not cyclically replaced as in elephants. Deinotheres reached peak abundance in the late Miocene; the largest species was *Deinotherium giganteum*, which reached about 4 m at the shoulder.

Palaeomastodonts and mastodonts

From the early Oligocene of North Africa we know palaeomastodonts. Some of these were the size of small elephants. There are two distinct types of palaeomastodonts; some such as *Palaeomastodon beadnelli* had upper and lower tusks that were oval in section and cheek teeth that were zygodont, that is with cusps arranged transversely in pairs along the tooth. Others such as *Phiomia serridens* have lower tusks that were flattened and bunodont molars in which the cusps are arranged in staggered pairs. In both lineages there was a diastema due to the absence of anterior premolars. While they succeed the moeritheres they are probably not direct descendants. After the palaeomastodonts there is a gap of 5 million years or more before we see proboscideans again. Then in the earliest Miocene we know mastodont faunas from both Africa and the Indian subcontinent; slightly later they appear in western Europe and by mid Miocene times they have reached North America. They did not, however, invade South America until just before their extinction in the Pleistocene. Mastodonts remained the dominant megamammals almost worldwide throughout the Miocene, but were replaced during the Plio-Pleistocene by the elephants.

While remains of tusks, teeth and limb bones are very readily recognizable as mastodonts, their specific identification, and still more their generic attribution, is often bewilderingly difficult. In part this is due to having to deal with any one of six cheek teeth, upper or lower, worn or unworn, male or female. Students have tried to identify stable features amongst the polymorphic forms, but few agree on these. The term mastodont is used here in the broadest sense to include mammutids and gomphotheriids. There appears to be at any one time in any one region about three kinds of mastodont, each kind presumably adapted to living in a different ecological niche. Mastodonts were all essentially browsing mammals in woodland savannahs. It was during the Miocene that the grasses became established as a major floral component, but the mastodonts appear to have remained folivorous.

A well known and widespread early mastodont is *Gomphotherium*, found for example in France, Kenya and Pakistan. It was nearly as large as an Indian elephant, with a pair of tusks in both upper and lower jaws; the short neck and long limbs indicate that a fully developed trunk must have been present. The multicusped cheek teeth were low crowned with thick enamel, with at least three teeth functioning simultaneously in each jaw half. The name mastodont refers to the domed cusps which occur in pairs along the length of the teeth. About ten genera of mastodont are recognized currently from the Miocene deposits of Eurasia and Africa. *Mammut americanum*, which by a perversion of nomenclature is a mastodont and not a mammoth, lived on into the Pleistocene and a large bull from Ohio stood over 3 m at the shoulder; it has zygodont molars and massive inwardly curved tusks which are over 2.5 m long.

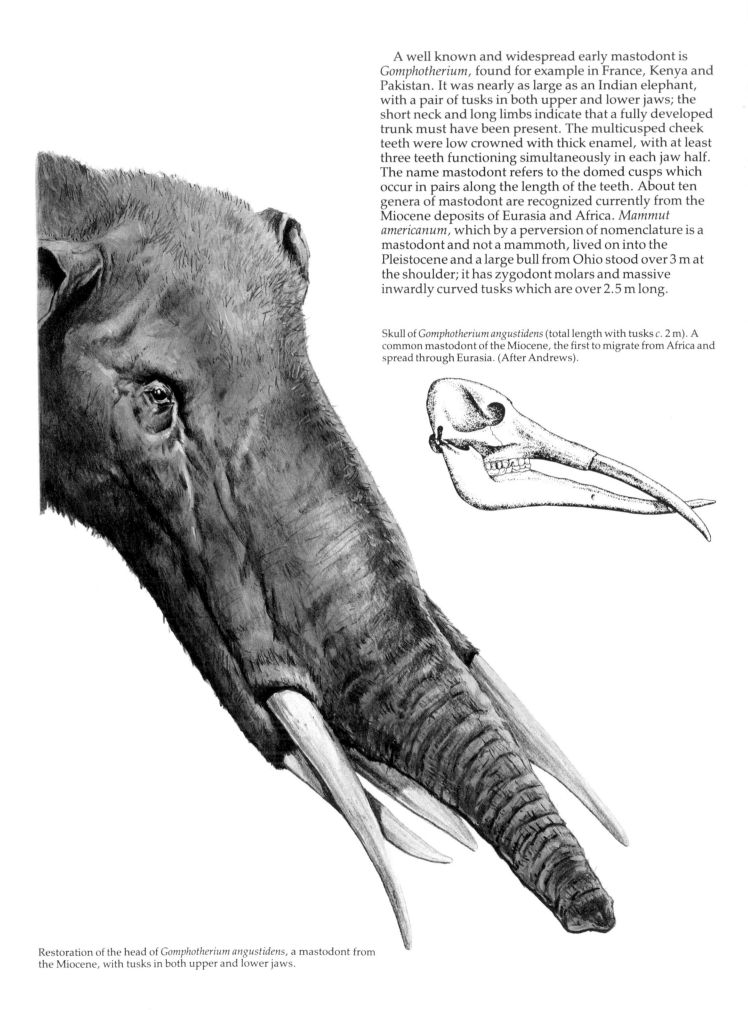

Skull of *Gomphotherium angustidens* (total length with tusks *c*. 2 m). A common mastodont of the Miocene, the first to migrate from Africa and spread through Eurasia. (After Andrews).

Restoration of the head of *Gomphotherium angustidens*, a mastodont from the Miocene, with tusks in both upper and lower jaws.

Restoration of *Ambelodon* from the latest Miocene of North America, a shovel tusked mastodont which used its mandible to uproot tubers.

Below: restoration of *Mammut americanum*, the American mastodont. This large proboscidean stood 3 metres at the shoulder, lived through the Plio-Pleistocene until about 10000 years ago and frequented spruce woodlands of North America.

Restoration of *Platybelodon*, a mid Miocene mastodont from Mongolia with the broad flat lower incisors forming a shovel. This mastodont is known from across Holarctica and from Africa.

Well known in the Asiatic Miocene sequences was *Platybelodon*. In this genus the upper tusks are small but the lower tusks became completely flattened and grew side by side to form a shovel, used in rooting out marshy vegetation. Another lineage included *Anancus* from the European Pliocene; it had extremely long straight tusks which reached 3 m in length.

Stegodonts from the Old World form a stratigraphic and morphologic bridge between mastodonts and elephants, though they are not regarded as being derived from mastodonts nor are they ancestral to elephants. *Stegodon ganesa* from the Pliocene of India is named after the elephant deity attendant on Siva; it had magnificent three-metre-long upper tusks, while in the mandible only vestiges of tusks remained. Not more than two cheek teeth were present at a time in each jaw half; the molars had up to 14 lophs or transverse ridges, each loph composed of a row of thickly enamelled cusps; traces of cement are present in the wide V-shaped valleys of the low crowned shearing dentition.

Elephants and mammoths

The last great wave of proboscidean evolution in the Plio–Pleistocene gave rise to the elephants and mammoths, both included within the family Elephantidae. When we look for the distinctions between the extinct mastodonts and elephantids we are limited to comparing osteological characters. The postcranial skeletons show only minor differences of detail. The differences in the skull architecture relate to dental changes and associated jaw mechanics. It is easy to list these differences but difficult to interpret them functionally. On the basis of their dentition elephants

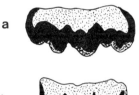

Vertical section through upper molar teeth of a series of proboscideans; all are orientated with occlusal or wear surface downward and anterior to the right, where wear begins. Enamel in black, cementum heavy stippled and dentine light stipple.

a *Gomphotherium*, a Miocene mastodont with three rows of low blunt cusps, thick enamel and a cementum capping.

b *Stegotetrabelodon*, a Pliocene elephant with three very widely spaced high conical lophs, deep V-shaped valleys and little cementum.

c *Primelephas*, a primitive Pliocene elephant. Increased number of cusp lophs, narrower V-shaped valleys, thinner enamel and more cementum in valleys.

d *Elephas planifrons*, early Pleistocene elephant. Tooth more high crowned, more lophs, narrower valleys, thinner enamel and more cementum.

e *Elephas maximus*, the living Indian elephant. Further development of features seen in *E. planifrons*.

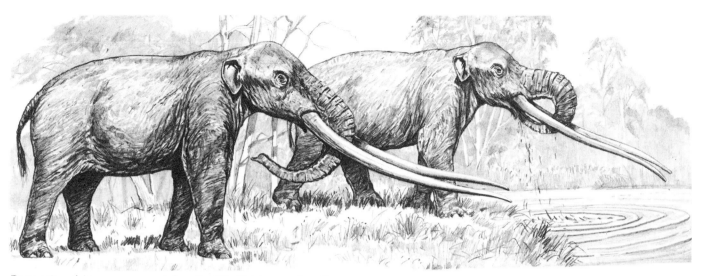

Restoration of *Anancus arvernensis* from the Plio-Pleistocene of southern France. This short-legged and long-faced mastodont has straight tusks reaching 3 metres in length.

might be regarded as primarily grazers – but this is known not to be so. In its natural habitat an African elephant is a browser on shrubs and trees, only resorting to grass when these fail; however, elephants differ radically from mastodonts in the way in which they process their food.

The emergence of elephantids from gomphotheriids took place in Africa in late Miocene times. An early distinctive feature is the great reduction and rapid loss of lower tusks and with this a shortening of the mandibular symphysis. This is of itself not unique to elephantids and can be seen in some mastodonts such as *Anancus* and the South American *Cuvieronius*. The loss of mandibular tusks enabled the elephants to change their mastication technique. In mastodonts grinding food was

accomplished by complex rotary movements. Elephants shear or cut their food with fore and aft jaw movements. To achieve a shearing action, the molar teeth developed a series of structural modifications. The cheek teeth have become completely cyclic with only one at a time in each jaw half. The number of lophs on each tooth increases, with up to 27 lophs or lamellae on the last molar of some mammoths; this results in a greatly increased length of enamel available for cutting. The lamellar frequency increases; that is, there are more lamellae in any given length of tooth, which enables the tooth to be more compact. There is also a decrease in enamel thickness; this helps reduce the bulk without decreasing the shearing length of surface, though at the cost of a more rapid wear rate. To compensate for this the enamel

Restoration of *Cuvieronius,* a mastodont from the Andes; this South American Pleistocene mastodont has spiral enamel on its tusks.

becomes crumpled, thus providing an increased length of shearing surface. Finally the teeth become very high crowned so that they take a long time to wear down. These dental trends are seen in the three lineages of elephantids that evolved from late Miocene ancestors. *Primelephas gomphotheroides* from Kenya is one such early elephant which still retained some mastodont dental characters; it has, for example, small mandibular tusks.

In Africa one lineage led to *Loxodonta africana*, the African elephant. During the Pleistocene the genus ranged over most of Africa, even across the Sahara which was sufficiently vegetated to support game; this lineage has retained many primitive features such as lozenge-shaped lamellae. The second lineage gave rise to *Elephas maximus*, the Indian elephant; it arose in Africa during the early Pliocene, radiating into Europe and Asia as far as Japan. In Pleistocene times several dwarf

Opposite: restoration of *Mammuthus imperator*, the largest mammoth known, reaching over 4 metres at the shoulder. Common in the late Pleistocene of North America from Alaska to Texas.

Restoration of *Stegodon ganesa* from the Pliocene of the Siwalik Hills in India. The long and heavy tusks are so closely placed there was no room between them for the trunk.

stocks appear in island faunas. From the European *Elephas namadicus* arose the dwarf *E. 'falconeri'* (it may be a variety or a species) which lived on the Mediterranean islands of Sardinia, Sicily, Malta, Crete and Cyprus; an adult would have stood less than one metre at the shoulder. From *E. planifrons* in Asia arose the dwarf *E. 'celebensis'* in Java and the Celebes.

The third lineage is that of the now extinct mammoths. These were the most specialized of all elephantids; the dental trends described above all reached their acme in the Pleistocene mammoths. The early history of mammoths is in Africa, whence they radiated in the early Pleistocene into Eurasia and North America. The woolly mammoth, *Mammuthus primigenius*, was abundant in the tundra regions of the Holarctic, reaching to the periglacial belt around southern Canada and north-east USA. The woolly mammoth is well known to us from completely frozen carcasses of animals that fell into crevasses. It was small (2·8 m at the shoulder), with high domed head, long and sloping back; the body was covered in long dense black hair with a thick undercoat and an 8-cm-thick fat layer to keep the cold out. The tusks have scars suggesting use in scraping snow off vegetation; they ate grasses and

Scale outline of two adult Pleistocene elephants. *Elephas antiquus* (shoulder height 3·7 m) was common through Europe; *E. falconeri* (shoulder height 0·9 m) lived on Mediterranean islands.

tundra legumes in summer, shrub leaves and bark in winter. The woolly mammoth survived into post-Pleistocene times, was hunted by Palaeolithic man and is vividly illustrated in cave paintings in France, dated around 20 000 years. In Siberia the mammoth died out only about 12 000 years ago.

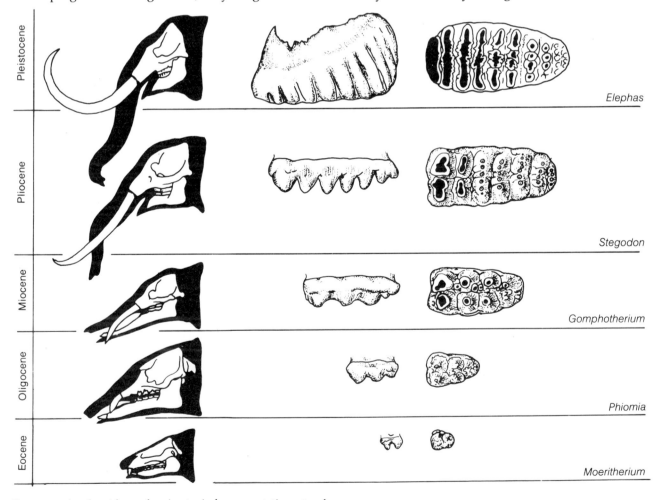

Sequence of proboscideans showing typical representatives at each time interval. The series does not form a direct phylogenetic lineage, but illustrates general morphological trends; these are increase in size, increase in tusk length, development of the trunk, increase in number of lophs or cusp rows on each tooth and increase in the height of the tooth crown. (After Lull and Thenius).

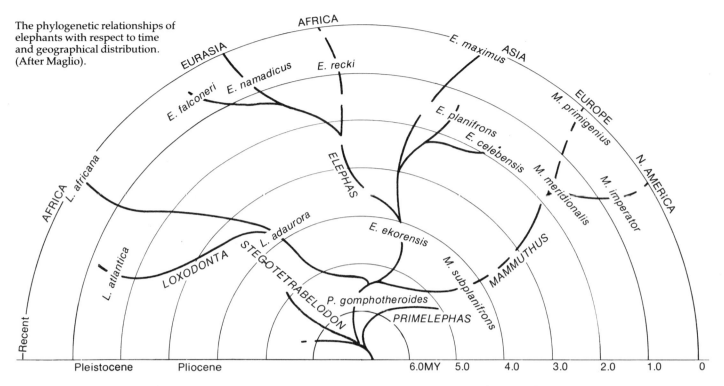

The phylogenetic relationships of elephants with respect to time and geographical distribution. (After Maglio).

In temperate zones of Eurasia the woodland straight tusked elephant (*Elephas antiquus* or *Elephas namadicus*) also survived the end of the last Ice Age, along with dwarf varieties (e.g. *Elephas falconeri*) on Mediterranean islands. In North America three other proboscideans survived the end of the Ice Age – the tundra woolly mammoth (*Mammuthus primigenius*), the woodland American mastodont (*Mammut americanum*) and the grazing mammoth (*Mammuthus jeffersoni*). Hunting by early man is the most likely cause of the final extinction of these great mammals, leaving only two survivors, the African and Indian elephants.

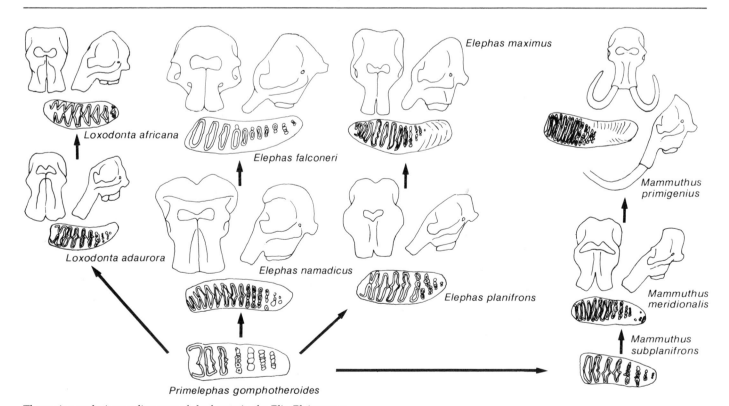

The major evolutionary lineages of elephants in the Plio-Pleistocene, showing the changes in molar pattern and skull proportion for each of the three genera (*Loxodonta, Elephas* and *Mammuthus*), all derivable from *Primelephas gomphotheroides* in the late Miocene of Africa. (After Maglio).

Restoration of *Mammuthus primigenius*, the woolly mammoth, which roamed the Holarctic regions during the last Ice Age, feeding on tundra vegetation.

CHAPTER 11

Mammals on island continents

Isolation is one of the essentials for natural selection to operate. The isolate may be a very small population, such as a tiny islet in a lake with a few mice; with the gene pool thus limited a new variety may arise in a short time. Island varieties of mainland species are common, e.g. rodents on Scottish islands. The stability and success of the new varieties depend on the length of time it takes for a new variate to become established, its inherent difference from the parental species and its viability in direct competition with the main population. Most varieties fail but a few make the grade.

The isolate may be on a larger scale, involving several species. During the Pleistocene, lowered sea-levels facilitated widespread dispersal of mammal species in the Mediterranean area. With the melting of the ice caps, sea-levels rose and thus created many islands, each possessing a handful of mammal species. These small populations adapted to the limited resources of their island habitats with size changes; the elephants and hippos became dwarfed.

Isolation also operates on a continental scale. The basic stocks present when isolation begins may be few in number and thus over tens of millions of years they have the opportunity to diversify into a host of ecological niches. The Caenozoic history of South America and Australia exemplify this pattern. When South America became isolated about 55 million years ago there were only about three basic mammal stocks on the continent; at the same time there were some 20 orders living in North America, which had land connections with Asia and Europe, and indirectly had links with Africa. Over the following 50 million years only two new terrestrial mammal stocks arrived in South America (primates and rodents). Australia became isolated perhaps 45 million years ago with only two basic mammal stocks, monotremes and marsupials. During the ensuing

millions of years the monotremes survived and the marsupials diversified, to be joined in the Pliocene by rodents, the first terrestrial placentals. It is the mammal history of these two continents that we survey in this chapter.

South American herbivores

For some 50 or 60 million years, from Palaeocene until Pliocene times, the continent of South America was isolated by seas from all other landmasses. The only possible relief may have been brief transitory island chains to North America, West Africa or Antarctica, but these, even if present, may not have been sufficient to support mammalian migrations. The consequence is that the mammals in South America during the Tertiary underwent a great adaptive radiation in isolation, and herbivores made up a large part of that radiation.

There is good geological evidence that towards the end of the Mesozoic there were land connections between Patagonia and Antarctica, a break occurring in early Palaeogene with the full circumpolar oceanic current becoming established by mid Oligocene times. Also during the late Mesozoic the Atlantic was opening up and as Africa drifted eastward, the last links with South America were between Brazil and West Africa along what is now the Ceare Rise, St Paul's Rocks and the Sierra Leone Rise. The Atlantic waters had crossed north of this by mid Cretaceous times about 90 million years ago, but an island chain probably survived into early Tertiary times. Between Central and South America the possible land connections are more difficult to reconstruct due to the complexity of the plate structures. We are not sure if in late Mesozoic times there was land, island chains or open sea between the two American continents. Certainly by early Tertiary it was essentially an open sea and this persisted

Maps to illustrate the palaeogeography of South America.

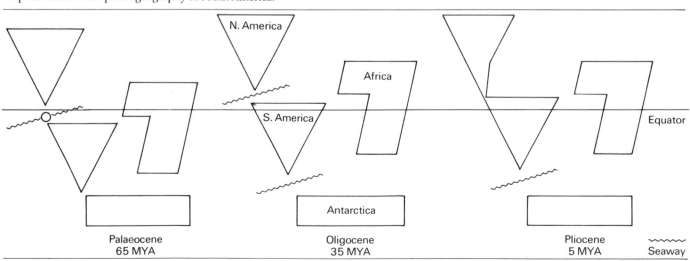

Palaeocene 65 MYA

Oligocene 35 MYA

Pliocene 5 MYA

Seaway

N. America

S. America

Africa

Antarctica

Equator

throughout most of the Tertiary. Two island chains became established during the early Pliocene, that of the West Indies and that across Central America and Panama, the latter later to become a firm land bridge.

The history of the South American mammalian faunas can be seen in three distinct phases. The first up to late Eocene times when three groups of mammals were present – marsupials, edentates and condylarths. These orders are also known on other continents and represent elements in the widespread distribution of late Mesozoic and early Caenozoic stocks. The second phase extends from the early Oligocene through to the late Miocene; marsupials and edentates radiate into many adaptive niches and the condylarths gave rise to a series of herbivores that diversified into six orders and about 25 families. To these are to be added two new stocks, primates and rodents which reached South America sometime before the early Oligocene by waif dispersal either from North America or more likely from West Africa. The third and final phase of the South American mammalian faunal history is spoken of as the Great Interchange when in Plio-Pleistocene times faunas migrated north and south across Central America, with South America acquiring about 15 northern families of mammals (mastodonts, horses, tapirs, peccaries, camels, deer, shrews, hares, squirrels, mice, dogs, bears, raccoons, otters and cats). Of these only the mastodonts have become extinct, but all the herbivores native to South America have sadly died out without leaving a single species; it is these mammals we will now discuss.

Litopterna

Didolodus is known from the early Eocene of Argentina and its dentition has close resemblances with that of the phenacodonts from North America. It is also a dentition that could serve as the archetype of litopterns and probably other South American herbivorous stocks. *Didolodus* is one of a handful of genera that form the family Didolodontidae and there are differing views on whether it should be grouped with the phenacodonts in the order Condylarthra or as a family within the order Litopterna.

Upper dentition of *Didolodus* (P^2–M^3 = 4·8 cm), from the early Eocene of Argentina. This primitive litoptern has six-cusped bunodont molars reminiscent of condylarths, from which it is derived. (After Simpson).

The didolodonts apart, there remain three established Litoptern families. The adianthids were small delicate creatures perhaps not unlike rabbits. The prototheriids constitute a major stock of horse-like animals ranging from late Palaeocene through to late Pliocene times. *Diadiaphorus* and *Proterotherium* had reduced their toes to

Restoration of *Diadiaphorus*, a pony-like litoptern from the Miocene of Argentina.

Thoatherium, top: skull (length 15 cm), skeleton (length 70 cm), and
hind foot (height 20 cm) (after Scott), and above: restoration. This fleet-
footed small antelope-like litoptern from the Santa Cruz beds
(Miocene) of Patagonia had a very short neck, short tail, unfused ulna/
radius but with fully one-toed foot, lacking even vestigial lateral toes.
The dentition was primitively low crowned with a very small
diastema.

Skeleton (length 1·2 m), skull (length 25 cm) and hind foot (height 35 cm) of *Diadiaphorus* from the Santa Cruz beds of Patagonia. The elongate slender legs retain unfused ulna/radius and tibia/fibula, but the toes are reduced to one functional (the third) digit and a pair of vestigial lateral toes. The dentition is almost complete, with only a short upper diastema resulting from the loss of the canine and two incisors. (After Scott).

SOUTH AMERICAN HERBIVORES

Order	Family	Generic examples
† Litopterna	† Didolodontidae	† *Didolodus*
	† Adianthidae	
	† Proterotheriidae	† *Diadiaphorus*
		† *Proterotherium*
		† *Thoatherium*
	† Macraucheniidae	† *Macrauchenia*
† Notoungulata		
† Notioprogonia		† *Arctostylops*
		† *Palaeostylops*
		† *Henricosbornia*
		† *Notostylops*
† Typotheria		† *Miocochilius*
		† *Mesotherium*
† Hegetotheria		† *Archaeohyrax*
		† *Pachyrukhos*
† Toxodonta	† Isotemnidae	† *Thomashuxleya*
	† Notohippidae	† *Notohippus*
	† Leontiniidae	† *Scarrittia*
	† Homalodotheriidae	† *Homalodotherium*
	† Toxodontidae	† *Toxodon*
		† *Mixotoxodon*
		† *Trigodon*
		† *Adinotherium*
† Astrapotheria		† *Astrapotherium*
† Pyrotheria		† *Pyrotherium*
Xenarthra	† Glyptodontidae	† *Glyptodon*
	† Megalonychidae	† *Hapalops*
		† *Pliometanastes*
		† *Nothrotherium*
		† *Nothrotheriops*
		† *Megalonyx*
	† Mylodontidae	† *Mylodon*
		† *Paramylodon*
	† Megatheriidae	† *Megatherium*
Pleistocene invaders		
	† Mastodontidae	Mastodonts
	Equidae	Horses
	Tapiridae	Tapirs
	Tayassuidae	Peccaries
	Camelidae	Llamas
	Cervidae	Deer

† Extinct

three, while in *Thoatherium* only the middle toe persists, the lateral digits being even more reduced than in the living horse; nevertheless the ulna and fibula though reduced did not fuse with the adjacent bone as in horses. Neither did the dentition parallel that of horses; the cheek teeth of prototheres are low crowned and not horse-like. *Thoatherium* was the smallest of litopterns and probably very gazelle-like in appearance. The prototheres had their acme in early Miocene times when they browsed the wooded plains of Patagonia and unlike their near relatives, the macrauchenids, they did not survive into the Pleistocene.

Brain cast of *Proterotherium*, which is relatively large though with little convolution of the cerebrum. (After Simpson).

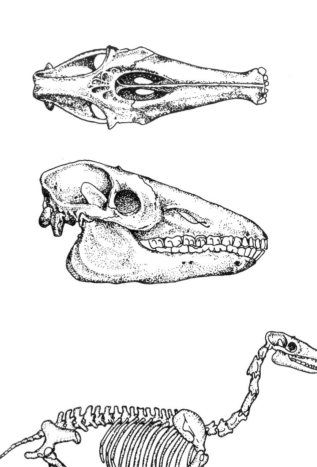

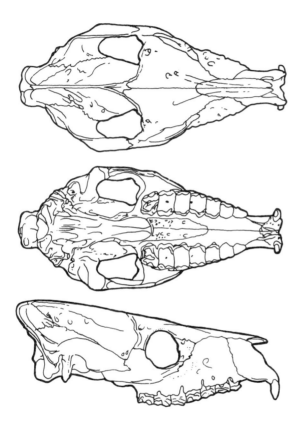

Skull of *Proterotherium* (length 18 cm) from the Miocene of Patagonia. There is a short diastema between the single incisor and the complete cheek dentition. (After Simpson).

Skeleton (length 3 m), skull (length 45 cm) and dentition of *Macrauchenia* from the Pleistocene of Patagonia, with three-toed feet, long neck, high nasal opening and complete dentition. (After Scott and Burmeister).

At Port St. Julian, in some red mud capping the gravel on the ninety-feet plain, I found half the skeleton of the Macrauchenia Patachonica, a remarkable quadruped, full as large as a camel. It belongs to the same division of the Pachydermata with the rhinoceros, tapir, and palaeotherium; but in structure of the bones of its long neck it shows a clear relation to the camel, or rather to the guanaco and llama.

So wrote Darwin of his discovery in Patagonia; the relationships of *Macrauchenia* (literally large camel) clearly puzzled him, with rhino-like feet, and camel-like neck. *Macrauchenia* also has a complete dentition of 44 teeth, a very unusual feature for a Pleistocene herbivore. The characteristically ridged upper molars and bicrescentic lower molars ally it with other litopterns, although the teeth are much more highly crowned. The cranium is short and the muzzle long with the nasal opening high up on the roof of the skull (as in sauropod dinosaurs). Despite the similarities of the three-toed feet to those of perissodactyls, the articular pattern of the ankle bones ('litoptern' means simple ankle) indicates affinities with notoungulates and condylarths, thus confirming the dental characters. Macraucheniids must have been very strange beasts – camel-like with a proboscis; they probably were both browsers and grazers on the Pampas plains but sadly did not survive the Pleistocene for us to see.

Notoungulata

Literally 'southern ungulates', the notoungulates consitute by far the largest order of South American herbivores, with four suborders, 14 families and well over 100 genera, ranging from late Palaeocene to Pleistocene times. They are not quite exclusively South American; from the late Palaeocene and early Eocene of the Rockies in Wyoming come partial dentitions of a rabbit-sized animal named *Arctostylops* which has been placed among the notoungulates. Better known from abundant dentitions in beds of similar age in Mongolia and China comes *Palaeostylops* and allied genera, again placed among the notoungulates; these mammals compare most closely with *Henricosbornia* and *Notostylops* from beds of similar age in Argentina. The simple but distinctive character of their dentitions allies these taxa in the suborder Notioprogonia; they demonstrate that the notoungulates had become widespread by early Tertiary times, but only flourished thereafter in the Neotropics.

Notoungulates for the most part have short skulls, flattened above with a broad brain case. The brains of early genera are of moderate size and do not display any significant changes during the course of the evolutionary history of the order. The structure of the notoungulate ear is unique and a highly distinctive feature of the order; the ossified tympanic bulla is inflated and there is a second large chamber formed in the squamosal (the epitympanic sinus) which communicates with the bulla by a canal. Additionally in many notoungulates a septum in the bulla separates a hypotympanic sinus

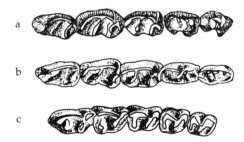

Lower dentitions of three notioprogonian notoungulates from three continents. (After Matthew and Granger).
a. *Palaeostylops* from the Palaeocene of Mongolia (1·3 cm);
b. *Arctostylops* from the early Eocene of Wyoming (1·7 cm);
c. *Notostylops* from the early Eocene of Argentina (4·5 cm).

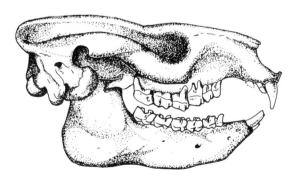

Skull of *Notostylops* (length 14 cm), a primitive notoungulate from the early Eocene of Patagonia. (After Simpson).

Restoration of *Macrauchenia*; camel-like, long necked and proboscis bearing with rhinoceros-like feet, this Pleistocene browser was the last of the litopterns.

area. This very specialized ear region strongly suggests acute hearing, perhaps similar to that known in desert living mammals today. In early notoungulates the dentition is complete with 44 teeth, but soon the canines and lateral incisors are lost to create a diastema. The anterior incisors are often chisel-shaped. The molars are superficially rhinoceros-like, the uppers with the three lophs arranged in a π shape; the lower molars are bicrescentic. The feet are mesaxonic, that is the third digit forms the axis, with some reduction of the lateral digits so that three is the usual number; the feet did not become truly unguligrade, most were hoofed but a few bore claws. The notoungulates had their acme in the Oligocene, were still abundant in the early Miocene and declined to extinction in the Pleistocene.

Besides the Notioprogonia, two other suborders of notoungulates contain small-sized mammals, the Typotheria and the Hegetotheria, both stocks making their appearance in the late Palaeocene and both persisting into the Pleistocene. They were abundant and varied, but few were larger than peccaries and they paralleled rodents in many ways, adapted to gnawing with chisel-like incisors followed by a diastema, and often with hypselodont cheek teeth (i.e. high crowned and open rooted). *Miocochilius* was a mid Miocene typothere with three toes on each front foot but with only two toes on each hind foot, the third and fourth digit being equal in size; this is very unusual for notoungulates in which the third or middle digit was normally used as the main weight bearer. *Mesotherium* was the last typothere in the Pleistocene; it had five nailed toes on each foot and a beaver-like skull in which chisel-shaped incisors are followed after a diastema by rootless hypselodont cheek teeth. Among the hegetotheres we may mention *Archaeohyrax*, which literally means ancient hyrax; it lived during Oligocene times in Argentina and Bolivia, and it looked very like a hyrax. But this resemblance is due to convergent

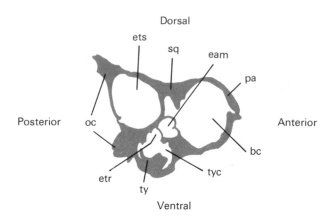

Longitudinal section through the ear region of the notoungulate *Oldfieldthomasia*, to illustrate the expanded bony bulla, tympanic and the epitympanic sinus. (After Simpson). bc brain cavity; eam external auditory meatus; etr epitympanic recess; ets epitympanic sinus; oc occipital condyle; pa parietal bone; sq squamosal bone; ty tympanic bone; tyc tympanic cavity.

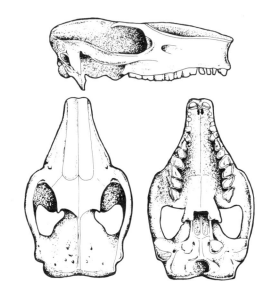

Restoration and skull (length 14 cm) of the coney-like *Archaeohyrax*, a hegetothere from the Oligocene of Patagonia. (After Ameghino).

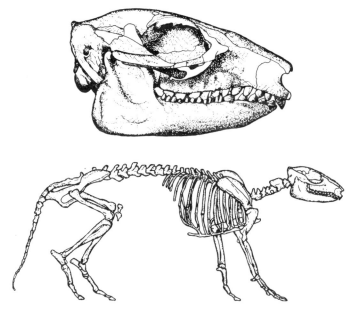

Skeleton (head and body 70 cm) and skull (length 15 cm) of the interathere *Miocochilius* from the Miocene of Colombia. (After Stirton).

Head and skull of *Mesotherium* (length 27 cm), a capybara-sized
Pleistocene typothere with a pair of very beaver-like incisor teeth.
(After Rovereto).

Restoration and skull (length 10 cm) of the rabbit-like *Protypotherium*
from the Santacrucian of Patagonia; the skull has a complete dentition
without any diastema. (After Sinclair).

evolution rather than ancestral affinity, as with so many of the Neotropical fauna. A Miocene hegetothere *Pachyrukhos* had hind limbs considerably longer than its forelimbs and almost certainly loped along just like a rabbit.

The final suborder, Toxodonta, contains many spectacular and large mammals; they comprised the main herbivore stocks on the continent throughout the Oligocene, Miocene and Pliocene. During the Pleistocene the genus *Mixotoxodon* reached Guatemala in Central America, but soon after all had become extinct. Toxodont means bow tooth and refers to the sideways curve on the cheek teeth. The dentition in

early genera was low crowned and complete, but in later forms a diastema appeared separating the anterior cropping teeth from the cementum sheathed and high crowned cheek teeth in which the simple pattern of the ridges is little different from that of early types. The ulna is always free but the fibula may fuse proximally with the tibia. Three-toed hoofed feet are normal.

Of the five families of Toxodonta, the isotemnids are the most primitive. A well-known genus is *Thomashuxleya*, named after the famous Victorian naturalist Thomas Huxley. This is only one of many such eponymic genera devised by the great Argentinian palaeontologist Florentino Ameghino; others include *Oldfieldthomasia, Ricardolydekkeria, Guilielmofloweria, Asmithwoodwardia, Carolodarwinia* and *Ricardowenia*, the latter two in particular having contributed much to South American mammal studies. *Thomashuxleya* from the early Eocene of Argentina has a complete dentition and the canines form small tusks. The skull is large in proportion to the body, which was about the size of a warthog; the animal stood digitigrade on its toes which were hoofed.

The notohippids might be renamed pseudohippids, for *noto* means southern and *pseudo* means false; notohippids are southern but are not related to horses.

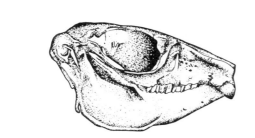

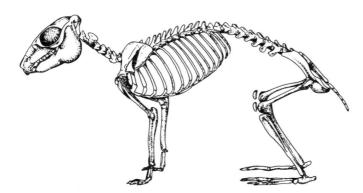

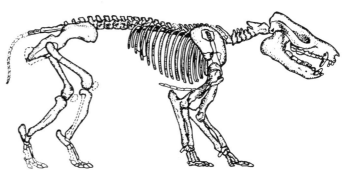

Skull (length 7 cm) and skeleton (length 30 cm) of *Pachyrukhos* from the Miocene of Patagonia. Note the rabbit-like proportions of this hegetothere with pointed incisors and a diastema behind, and the long hind legs. (After Scott).

Restoration and skeleton of *Thomashuxleya* (length 1·3 m), a 50-million-year-old warthog-like toxodont with digitigrade hoofed feet from Patagonia. The skull is proportionately large and the dentition complete. (After Simpson).

The skull and cropping incisors vaguely recall that of horses, but everything else is as in notoungulates. Already in late Eocene times notohippids had very high crowned cheek teeth with a cementum sheath, features not seen in true horses for another 25 million years. *Notohippus*, the last member of the family in the early Miocene, would have been able to exploit the new grassy plains.

The leontiniids are a small family in which *Scarrittia* is the best known genus. This beast was about the size of a tapir, with short face, complete dentition with low crowned teeth, very short tail and the tibia and fibula fused at the upper end so that the three-toed hoofed foot could not be turned sideways.

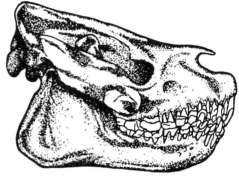

Skull of *Scarrittia* (lengtn 48 cm); the short face accommodates a complete dentition with incisiform first upper incisor. (After Chaffee).

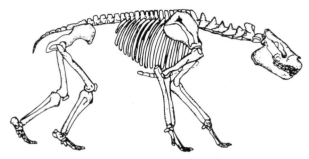

Reconstruction and skeleton (length 2 m) of *Scarrittia* from the Deseadan. (After Chaffee).

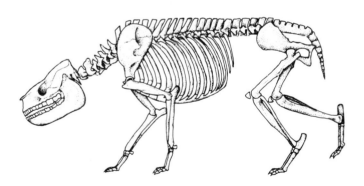

Skeleton (length 1 m), upper dentition (I^1–M^3 = 8 cm) and restoration of *Rhynchippus*, a 35-million-year-old notohippid from Patagonia. The incisors are horse-like and the high crowned molars rhinoceros-like. (After Simpson). It was a three-toed toxodont with clawed feet. (After Loomis).

Homalodotherium from the middle Miocene is the best known homalodotheriid. It was about the size of a llama and differed from all other notoungulates in having claws on each of its four front digits. The forelimbs were long and heavily built and the hind limbs relatively short. The teeth were high crowned, rooted and without cementum. *Homalodotherium* is thought to have been a bipedal browser, using its clawed arm to achieve greater reach in obtaining foliage. In these features it closely resembled the chalicotheres of North America and the Old World.

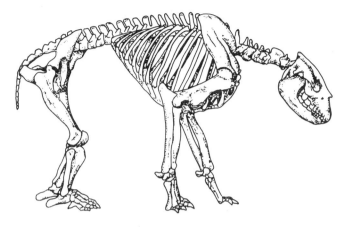

Restoration and skeleton of *Homalodotherium* (length 2 m), a 20-million-year-old llama-sized toxodont from the Santacrucian of Patagonia. It was a bipedal browser with robust forelimbs and short hind limbs, each with four clawed toes. (After Riggs).

It was Darwin who found the first specimens of toxodontids in the Pampean loess and he wrote:

'Toxodon, perhaps one of the strangest animals ever discovered. In size it equalled an elephant or megatherium; but the structure of its teeth, as Mr. Owen states, proves indisputably that it was intimately related to the Gnawers, the order which at the present day includes most of the smallest quadrupeds. In many details it is allied to the Pachydermata. Judging from the position of its eyes, ears, and nostrils, it was probably aquatic, like the dugong and manatee, to which it is also allied. How wonderfully are the different orders, at the present time so well separated, blended together in different points of the structure of the toxodon!'

So puzzling was the beast that Professor Owen in his account of Darwin's specimens created for them a new order Toxodontia (equivalent of the present suborder Toxodonta). *Toxodon* exemplifies the big curved or bowed cheek teeth which are very high crowned with open roots (hypselodont) so that they could grow throughout the life of the animal to compensate for the heavy wear on the tough Pampas grazing. There was a diastema, chisel-like upper incisors and procumbent lower incisors. The high nares suggest a prehensile lip may have been present. In all these features the skull is very rhinoceros-like. The ear was placed very high on the skull as in other notoungulates. The neck was short and the pelvis broad as in graviportal mammals. The limbs were short and massive, hind longer than front and the tibia and fibula fused proximally. The feet were three-toed, hoofed and plantigrade, again similar to that in rhinos and elephants. In total the nearest living analogue is a rhinoceros; a large, heavy, slow moving mixed feeder on foliage and grasses.

Skull of *Homalodotherium* (length 45 cm) showing complete dentition without a diastema and rhinoceros-like pattern on the high crowned molars. (After Patterson).

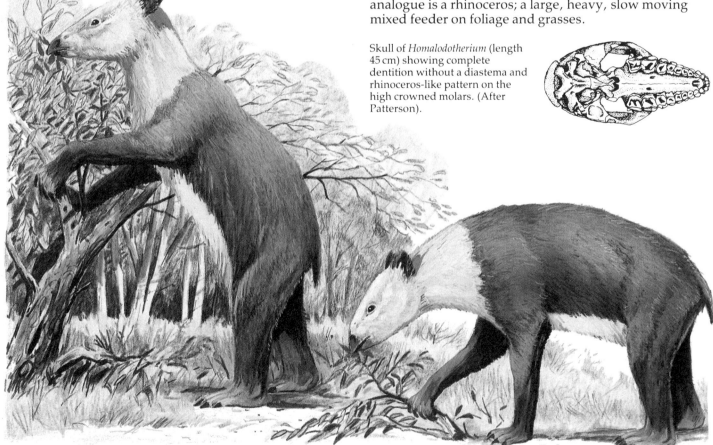

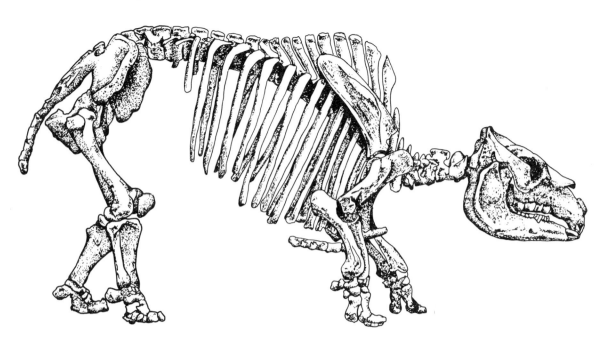

Skeleton (length 2·7 m), and restoration of *Toxodon* from the
Pleistocene. This graviportal, rhinoceros-like notoungulate from the
Pampean of Argentina has hind limbs which are longer than the front
limbs, fused tibia/fibula and three-toed hoofed plantigrade feet. The
skull has a hippo-like constriction behind the chisel-edged incisors.
(After Piveteau).

Other toxodontids include *Trigodon* which had a horn on its forehead as do living rhinoceroses. Some toxodontids such as *Adinotherium ovinum* were, as the name suggests, small sheep-sized, and in the males a small frontal horn was present.

Head of *Trigodon*, Pliocene toxodont with a frontal horn.

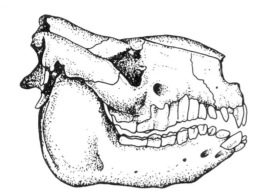

Nesodon, above: skull (length 40 cm) and upper cheek dentition. (After Scott). Below: restoration of this Santacrucian toxodont from the mid Miocene of Patagonia, with high shoulders and head slung low on long neck.

Restoration of *Adinotherium*, a 20-million-year-old toxodont with small frontal horn.

Astrapotheria

Astrapotherium from the famous Santa Cruz volcanic ash beds of early Miocene age in Argentina is the best known member of this small and enigmatic order of mammals. Astrapotheres are strange beasts even by South American standards; we know almost complete skeletons and still do not understand how they lived. *Astrapotherium* was as large as a rhinoceros, with a long body, and short feeble legs and small feet. The skull was very short and high nasal bones suggest the presence of a proboscis, yet the neck is not shortened and the head could easily have reached the ground. The forehead was domed by air sinuses. There were no upper incisors but the lower ones were broad and procumbent and possibly used against a hard pad in the upper jaw to crop vegetation. The canines formed tusks, the upper large and vertical shearing against the smaller lower tusk rather as in hippos. Behind a large diastema the cheek teeth were hypsodont and not dissimilar to those of notoungulates. The strange combination of features suggests perhaps a semi-amphibious mode of life, not unlike the extinct amynodont rhinoceroses of the northern hemisphere.

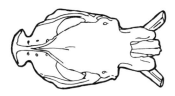 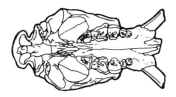 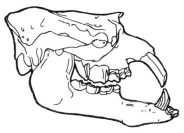

Restoration, skull and skeleton of *Astrapotherium*, a 20-million-year-old rhinoceros-sized ungulate from Patagonia with a short proboscis. The skeleton (length 2·5 m) has a long neck, short legs and small feet. (After Scott). The large skull (length 50 cm) has high nasal bones and large shearing hippo-like canines. (After Simpson).

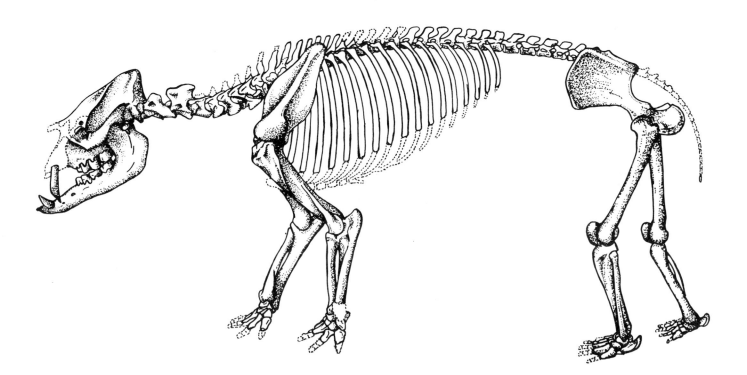

Pyrotheria

This final order of strange Neotropical ungulates ranged from Argentina through Brazil to Venezuela and Colombia during the Eocene and early Oligocene. *Pyrotherium*, literally 'fire beast', is so named from the volcanic ashes of Deseado where the levels containing their bones are known as the Pyrotherium Beds. *Pyrotherium* was a large animal, the size of a small elephant. The skull had two upper and one lower incisor forming well-developed tusks and, behind a large diastema, a set of bilophodont cheek teeth. The high nares suggest the presence of a proboscis; the postcranial remains are few but those that we know are distinctly elephantine in appearance. *Pyrotherium* has a very close resemblance to its African contemporary *Barytherium*, which is grouped with the proboscideans. However, *Pyrotherium* is believed to have a truly notoungulate ear region, which would indicate yet another case of incredibly close convergence.

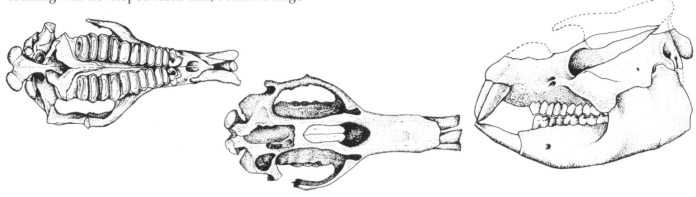

Pyrotherium, top: skull (length 85 cm) and, above: restoration of the head and fore quarter. From the volcanoclastic deposits of Patagonia this 35-million-year-old small elephant-sized mammal has hippo-like anterior dentition and *Barytherium*-like molars. (After Loomis).

Xenarthra

Of all strange beasts in the fossil record, the Xenarthra are the strangest. Their many varied members include the folivorous tree sloths which are almost without a fossil record, the insectivorous anteaters and armadillos, together with two major herbivorous and extinct stocks, the glyptodonts and ground sloths; these two fossil groups were restricted to South America until Miocene times when some began to colonize the Antillean island chain and from there moved into North America.

Most xenarthrans possess extra vertebral articulations – hence their name 'strange joints'; the feet usually bear large claws, some genera had ossicles in the skin, others had a bony shield over the body. The living xenarthrans have a variable body temperature and one which is lower than that of other placental mammals. In very many ways, xenarthrans are the most bizarre of mammals.

GLYPTODONTIDS   This group evolved from armadillo-like ancestors and we can trace their history through about 50 genera from early Miocene times when they were medium-sized mammals with a flexible shield made up of a mosaic of ossifications. The family achieved its acme with the great Plio-Pleistocene forms that grazed the pampas of Argentina and reached the southern states of America. *Glyptodon* was gigantic, standing 1·5 m in height and 3·3 m in length; as large as a small motorcar and armoured like a tank. The rigid bony carapace enclosed all the body except the underside and was made of a mosaic of sculpted

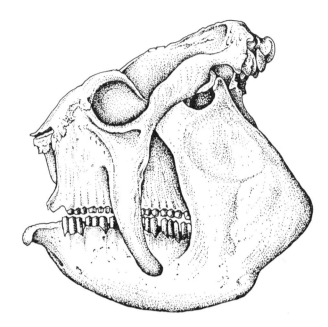

Glyptodon, above; skull (length 30 cm) from the Pleistocene of Argentina. The massive deep jaw and elongate zygomatic process served for attachment of heavy muscles used in masticating tough vegetation. (After Burmeister). Below: restoration showing the full armour of bony helmet, body carapace and spiked tail.

polygonal bony scutes; in addition there was a bony helmet protecting the head and it had an armoured tail to lash any would-be assailants. It must have been quite impregnable inside the shield, which alone weighed around half a tonne. Allowing 2 tonnes for the weight of the animal, this means that about 20 per cent of the total weight was bony carapace – an enormous burden to carry. The tusks of an elephant are a mere three per cent and the antlers of a Giant Irish Deer about 14 per cent of the total body weight; however, a fully loaded camel can carry an incredible one third of its own weight.

The head of *Glyptodon* has deep jaws within which are set open rooted columnar cheek teeth, each with three

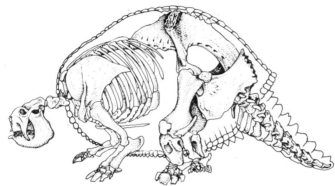

Skeleton of *Glyptodon* (length of head and body 2 m). Note the enormously heavy pelvic girdle and femora, short clawed limbs and massive tail, the whole skeleton enclosed in a bony carapace. (After Burmeister).

Three of the thick polygonal ornamented bony plates that make up the mosaic composition of the rigid heavy carapace in *Glyptodon*. (After Owen).

Upper dentition of *Panochthus* (length 22 cm), a Pleistocene glyptodont from Patagonia with characteristically eight columnar three-ridged cheek teeth. (After Burmeister).

lobes, thus giving the appearance of a carved tooth and hence the name *Glyptodon*; with this dental battery it could have obtained nourishment from a variety of abrasive plants, including grasses. The pelvis is very large and attached at two points to the fused posterior vertebrae; the hind limb is very massive and the feet have large blunt hoof-like claws. Their frequent association with capybaras suggests a preference for aquatic vegetation and they may have sported a short proboscis.

GROUND SLOTHS   The second major group of extinct xenarthrans are the ground sloths which range in time from the early Oligocene up to the end of the Pleistocene and even beyond. Ground sloths, like glyptodonts, have no living analogues; they were bipedal browsers, diversifying into about 60 genera ranging from cat-sized species to giants over six metres long. The fossil record begins in the early Oligocene and already by that time three families of sloths had become established (megalonychids, mylodontids and megatheriids). The early megalonychids such as *Hapalops* were small semi-arboreal animals. *Pliometanastes* was the first North American ground sloth in the early late Miocene of Florida about 10 million years ago, reaching North America via the Antilles. *Nothrotherium* was common in the Pleistocene of South America, as was its big brother *Nothrotheriops* in North America; the latter is one of the best known ground sloths as skeletons with muscle and skin are preserved in the dry caves of the southwestern states of America. It was 2·5 m long and 1 m high and weighed around 160 kg. It had yellowish hair which was tinged green due to the growth of algae in the fur, as happens on living tree sloths. The coprolites (fossil dung) of *Nothrotheriops* show that it fed on a diet of desert plants such as *Ephedra* and *Yucca*, procured with its spout-like mandible. It probably wintered in the caves and spent the summer higher up the mountains. The 'freshest' coprolites dated show it to have been still living 11 000 years ago.

On 10 March, 1797, Thomas Jefferson gave a talk to the American Philosophical Society in which he described a 'large clawed' beast, *Megalonyx*, found in a cave in West Virginia; the name refers to the great claw on the third digit on the hind foot. The lecture is often taken as marking the beginning of vertebrate palaeontology in North America. This Pleistocene megalonychid represents the acme in the evolution of the ground sloth family; it was very large though not as big as *Megatherium* (see below). Besides the usual battery of grinding cheek teeth it had a pair of anterior caniniform cutting teeth that grew persistently. There appear to have been three distinct species in North America, probably feeding on different foliage in the woodlands. *Megalonyx* had a very widespread range extending over much of America, through Canada to Alaska; it survived until the end of the Pleistocene in North America. Also during the Pleistocene there arose several dwarf cat-sized genera on islands of the West Indies.

Mylodontids are intermediate in size between megalonychids and megatheriids. *Mylodon* survived in Patagonia until Recent times, the last individuals being

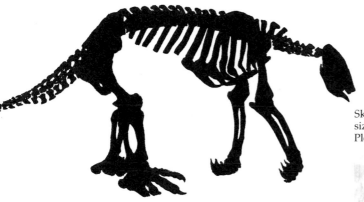

Skeletons of *Hapalops* and *Nothriotheriops* to same scale, illustrating the size increase over 25 million years between the Miocene and the Pleistocene.

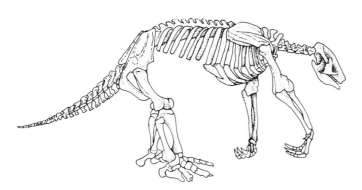

Skeleton of *Nothriotheriops* (head and body length 1·6 m); a large megalonychid ground sloth from Rancho la Brea, California. (After Scott).

Restoration of *Mylodon*, the last surviving ground sloth from a recent deposit in Patagonia.

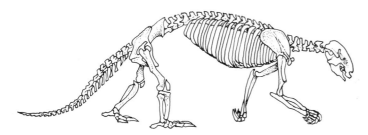

Skeleton of *Hapalops* (head and body length 70 cm); a small semi-arboreal megalonychid ground sloth from the Santacrucian of Patagonia. (After Matthew).

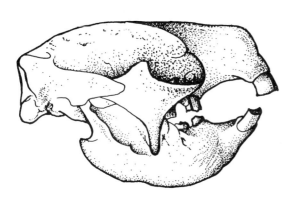

Skull of *Megalonyx* (length 66 cm), a very large ground sloth that was widespread in North America during the Pleistocene. (After Scott).

Restoration of *Nothrotherium*, a South American Pleistocene megalonychid ground sloth.

Skeleton of *Glossotherium* (head and body length 3 m); a large mylodontid ground sloth from the tar pits of Rancho la Brea, California. (After Stock).

killed by man. In caves we can still find the red haired skin in which are buried bony ossicles. A similar ground sloth *Glossotherium* has been recovered from the late Pleistocene tar pits of Rancho La Brea in California.

In September, 1789, there arrived in Madrid an almost complete fossilized skeleton of a giant beast for the Royal Cabinet. It had been sent by the governor of Buenos Ayres and was a few years later described by the great French palaeontologist Cuvier as a giant sloth under the name of *Megatherium americanum* – the gigantic American beast. It is the largest ground sloth, six metres from head to tail and probably over three tonnes in weight. The skull has an extremely deep zygoma for the attachment of a very powerful masseter chewing muscle; the cheek teeth are squarish in section with two transverse crests, high crowned and cement lined. It has large forelimbs and short but massive hind limbs. The toes were clawed, the third digit on the hind foot having an especially large claw; the foot structure necessitated the animal walking on the outer edge of its foot. *Megatherium* was probably a browser and a grazer in grassy woodlands or wooded plains, and additionally it may have frequented dry pastures where with its great claw it could uproot tubers, perhaps to obtain their store of drinking water. The family ranged widely through South and Central America and up into the southern states of USA.

Pleistocene invaders

To conclude we will glance briefly at the herbivores that invaded South America about three million years ago and in less than two million years had become established over much of the continent while the native stocks declined to extinction. Of the six invading families only the mastodonts have become totally extinct. These reached South America in the early Pleistocene and rapidly became abundant on the continent; at least two distinct genera are recognized – *Stegomastodon* with curved tusks from, for example, the Pampas of Argentina and *Cuvieronius* with straight tusks, well known from the Southern Andes. The horses have a similar history of invasion, diversification into two genera and extinction by the close of the Pleistocene; horses also became extinct in Central and North America, but survived in the Old World. The cause of

the total extinction of horses in the Americas is unknown; perhaps it was due to an epidemic disease sweeping through the populations, but this and other hypotheses are untestable.

The tapirs inhabit Central and South America and south-east Asia, but have been extinct in North America since the Pleistocene; they are basically tropical moisture loving animals. The peccaries reached South America earlier than the other invaders, in late Pliocene times. Peccaries are rather pig-like animals, though not very closely related to pigs, and today they range from Texas to Patagonia. Llamas (members of the camel family), have been particularly successful invaders though they have become extinct in North America. They have adapted well to the steppe and cold climate of the high Andes. Two species have been domesticated (alpaca and llama) and are now unknown in the wild; the guanaco and vicuña are still to be found wild, though rare, and at altitudes over 5000 metres in the Andes. The deer, with the peccaries, are the only family of invaders that today inhabit both continents (excepting the horses reintroduced by man). There are six genera of deer in South America; four are endemic to the continent. Of these *Blastocerus* the swamp deer is the biggest (100 kg) and *Pudu* the smallest (10 kg). The Virginia deer *Odocoileus* has a range from Canada to northern South America and *Mazama* the brocket deer ranges through tropical South and Central America.

The invasion of the South was not without a counter invasion into North America, though in this case only eight mammal families made the journey. Looking at the herbivores, the notoungulates did not get beyond Central America, but four families of glyptodonts and ground sloths successfully penetrated and flourished on the North American continent.

During the course of the Caenozoic, changes in the climate and geography of South America are reflected in the mammalian history. Our picture in Palaeocene and Eocene times is of tropical rain forests in Brazil and woodlands in Patagonia. As time progressed, the southern areas became cooler and drier. Desertification ensued as the Andes rose and a great rain shadow belt was established in Patagonia. The increasing aridity caused the woodland to give way first to savannah and later to steppe. By Miocene times we see a dramatic increase in the proportion of hypsodont herbivores, feeding on the tough grasses of the open steppe. The savannah faunas of the Oligocene and early Miocene were highly varied and abundant; by mid Miocene times the diversity was falling as the savannahs declined. Two-thirds of the invading mammals of the early Pleistocene were savannah adapted species with grazers predominating. These soon adapted to the steppe habitats of the Andes where the survivors are to be seen today. The cause of the extinction of the vast array of native South American mammals at the close of the Pleistocene is unknown, but among the factors influencing their decline must be the deteriorating climate, loss of savannah habitats and the competition with invaders.

Opposite: Restoration of *Megatherium*, the largest of the great Pleistocene ground sloths from Patagonia.

Australian herbivores

Our knowledge of Australian fossil mammal faunas is in its infancy compared with that for North America. Both continents have a similar number of families of living terrestrial mammals (25 placental in North America and 23 marsupial in Australia). North America has a fossil record of over 70 extinct families; as yet Australia has only five or six known extinct families; there must be many unknown, though the stratigraphic record is such that they may not be preserved.

Although the Gondwana supercontinent had begun to fragment in Jurassic times, Australia remained joined to Antarctica until late Eocene times. During the Oligocene the circum-Antarctic ocean current became established. With the isolation of Antarctica, the ice cap began to grow, the ocean temperatures to fall and a cold nutrient-rich deep water biota developed (the psychrosphere). New Zealand became isolated during the Cretaceous and has no native terrestrial mammals, nor have any fossil ones been found. Australia is the only continent on which all three major stocks of living mammals (monotremes, marsupials and placentals) are to be found. The early fossil record is unhappily poorly known. Recently a marsupial fossil was found in Eocene strata in Antarctica, in an area close to South America; it demonstrates that mammals did live on the Antarctic continent in early Tertiary times and therefore dispersal could have occurred between South America and Australia via Antarctica, across the now icy continent.

Near Koonwarra in Victoria, Australia, there are freshwater lake deposits of early Cretaceous age which have yielded fish, bird feathers and some apparently 'mammalian' fleas; this tantalizing evidence of Mesozoic mammals has very recently been transformed by the discovery in New South Wales of a monotreme in beds of similar age. After this 100-million-year-old specimen, the next oldest fossil mammal is a marsupial a mere 25 million years old. By mid Miocene times, about 15 million years ago, we can recognize 13 families of marsupials, a monotreme and a bat; most of the marsupials are arboreal and represent browsing temperate rain forest faunas. By late Miocene times the scene changes; with drier conditions grasslands spread and we then see the rise of the grazing kangaroos. In early Pliocene times the rodents arrived in northern Australia from south-east Asia, presumably by waif dispersal.

Bills and quills – the monotremes

The egg-laying mammals – the billed platypus and the quilled echidnas – are survivors of a very early mammalian radiation, much more primitive than any marsupial or placental and clearly distinct from their reptilian ancestors. Though unknown on other continents, the monotremes appear to have their nearest relatives among the multituberculates, which had a widespread distribution across northern continents in Jurassic and Cretaceous times. Unhappily the Australian fossil record for monotremes is extremely fragmentary; an early Cretaceous fragment, a mid Miocene genus *Obdurodon* with much similarity to the living *Ornithorhynchus*, and some Plio-Pleistocene specimens. *Obdurodon* possessed enamelled cheek teeth, whereas in the living platypus the adults have rubbery teeth which are devoid of enamel and dentine. Both living genera of monotremes are extremely specialized; the platypus *Ornithorhynchus* is semi-aquatic inhabiting streams and lakes and using its bill to feed on the bottom fauna. The echidna *Tachyglossus* is hedgehog-like with spines and feeds mainly on ants and termites.

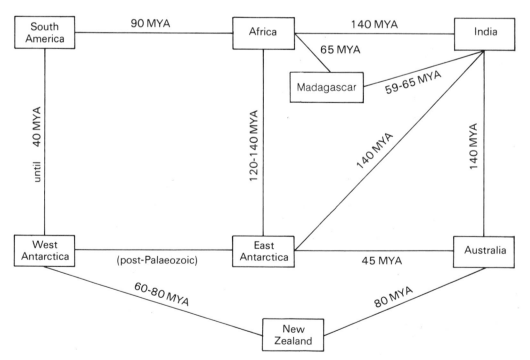

Diagrammatic representation of the approximate times of separation of Australia from other southern continents. (After Archer).

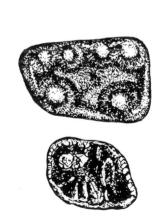

Cheek tooth of the Miocene monotreme *Obdurodon* (top) above that of the living platypus.

AUSTRALIAN MAMMALS

Monotremata

Marsupialia

Order	Family
Dasyurida	Dasyuridae
	† Thylacinidae
	Myrmecobiidae
Peramelina	Peramelidae
	Notoryctidae
Diprotodonta	Vombatidae
	Phascolarctidae
	† Thylacoleonidae
	† Wynyardiidae
	† Palorchestidae
	† Diprotodontidae
	Potoroidae
	Macropodidae

Generic examples
† *Obdurodon*
Ornithorhynchus
Tachyglossus
Zaglossus

† *Ankotarinja*
Sarcophilus
† *Thylacinus*
Myrmecobius
Perameles
Notoryctes
Vombatus
† *Rhizophascolonus*
¹ *Phascolonus*
† *Thylacoleo*
† *Wynyardia*
† *Ngapakaldia*
† *Palorchestes*
† *Neohelos*
† *Zygomaturus*
† *Eurzygoma*
† *Diprotodon*
Hypsiprymnodon
Bettongia
Dendrolagus
† *Prionotemnus*
† *Sthenurus*
† *Procoptodon*

† Extinct

Partial skeleton of the giant Pleistocene echidna *Zaglossus* (length of head and body about 65 cm). (After Archer).

Pouches for protection – the marsupials

As well as the distinctive differences in reproductive strategies, marsupials can be distinguished from placental mammals on a number of osteological characters. These include the presence of a pair of marsupial bones attached anteriorly to the pubis, an inflected angular process on the mandible, the presence of four molar and three premolar teeth and as many as five upper incisor teeth. The upper molars often have an outer row of stylar cusps and there is almost no tooth replacement; the last premolar is normally the only tooth with a milk precursor. There is no bony auditory bulla nor any postorbital bar. The front feet usually have five clawed toes and the hind feet often reduce the number of toes, the fourth being the largest clawed digit.

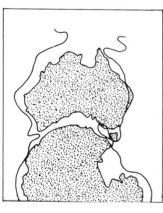
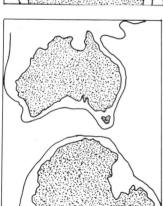

Palaeogeographic reconstruction of Australia 60 million years ago before the break with Antarctica and 45 million years ago after the break. The approximate coastlines at each interval indicated with thin line. (After Kemp).

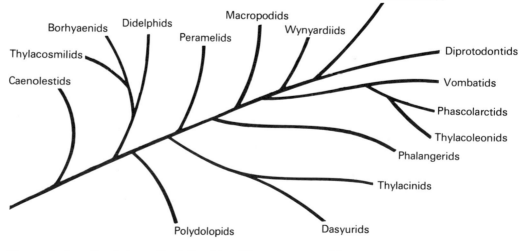

Diagram to show the phylogenetic relationships of the major marsupial families. (After Archer).

EARLY AND NON-HERBIVOROUS TYPES   The radiation of
Australian marsupial mammals is presumed to have
originated from a didelphoid stock comparable to the
American didelphids. The earliest Australian
marsupials probably comprised small insectivorous-
omnivorous mammals, not unlike the smaller living
dasyurids, which can be traced to the mid Miocene in
South Australia; here *Ankotarinja* is morphologically
intermediate between a didelphid and a dasyurid. The
larger dasyurids have become scavengers and
carnivores, as with *Sarcophilus*, the Tasmanian devil.
Two specialized sidelines of dasyurids produced
Myrmecobius, the banded anteater and the carnivore
Thylacinus, the Tasmanian 'wolf'; the latter is traceable
to the late Miocene in Northern Territory. *Thylacoleo*, the
marsupial 'lion' is described on page 70; it belongs to a
lineage traceable to the mid Miocene. The small
terrestrial peramelids (bandicoots), the arboreal
phalangeroids (possums, gliders and cuscus) and the
fossorial *Notoryctes* (pouched mole) form three important
radiations into a variety of ecological niches. All have a
poor or non-existent fossil record in pre-Pleistocene
deposits. This leaves the bulk of the known fossil faunas
of Australian mammals comprising the herbivorous
marsupials, both browsers and grazers.

BROWSERS AND GRAZERS   Apart from a late Oligocene
site with some unidentifiable mammal fragments, the
earliest Tertiary mammal site is an early Miocene locality
in northern Tasmania; this has yielded the skeleton of
Wynyardia. While the skull is preserved, unhappily the
teeth are not, and so its relationship to other marsupials
remains obscure.

Mid Miocene faunas are known from many localities
with representatives of 15 families; amongst the
herbivorous marsupials, browsers are dominant. These
herbivores comprise phascolarctids (koalas), vombatids
(wombats), macropodids (kangaroos) and two extinct
families, palorchestids and diprotodontids; all but the
macropodids were browsers. A major ecological change
occurred in early Pliocene times with an explosive
radiation of the bipedal grazing wallabies and
kangaroos; this we may associate with the drier climatic
regimes and the spread of grasslands.

The living koala is arboreal and tailless, and probably
atypical of phascolarctids; it appears to represent a
specialized sideline of a wombat-like stock. The
wombats have a relatively poor fossil record; five extinct
genera and two living genera are known.
Rhizophascolonus from the late mid Miocene is a primitive
wombat with rooted teeth; all later wombats have high
crowned rootless teeth, the better capable of shredding
coarse vegetation. *Phascolonus gigas* was a gigantic
Pleistocene species; it was as large as a pigmy hippo,
with the stocky build of a wombat. It is unlikely that this
giant fossil wombat burrowed as do its living relatives.
Phascolonus has a pair of very large spatulate upper
incisors and a pair of smaller lower incisors. The four-
cusped molars wear to give a double-lozenged shaped
pattern.

Palorchestids were common in mid Miocene times
with *Ngapakaldia*, a sheep-sized browser. *Palorchestes*
was common from late Miocene through into Pleistocene

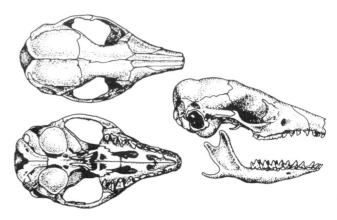

Skull and mandible of small dasyurid marsupial *Sminthopsis*. The
marsupial features include four upper incisors, three premolars and
four molars, and a projecting angular process on the mandible. (After
Archer).

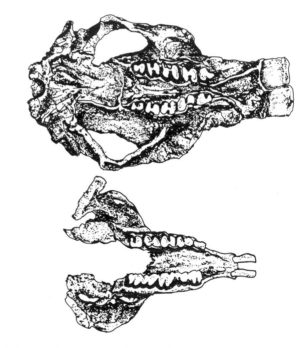

Skull (length 40 cm) of *Phascolonus*, a giant wombat from the
Pleistocene of southern Australia with a broad pair of chisel-shaped
upper incisors. (After Dawson).

An early Cretaceous flea from
Victoria, Australia. The specimen
is similar to fleas which currently
infest mammals. (After Riek).

Upper molar tooth of *Ankotarinja*,
a primitive Miocene dasyurid.

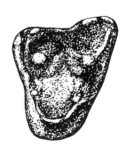

times; this large animal was as big as a bull, had powerful front limbs with large compressed claws, a short neck and short tail. The nasal bones are retracted and the rostrum narrow, giving the impression that a proboscis was present. These strange folivores were perhaps the marsupial analogues of the ground sloths. Along with other large marsupials, the palorchestids became extinct during Pleistocene times.

The diprotodontids are the sister group to the palorchestids and range from mid Miocene through into recent times; there is evidence that they were hunted by aboriginals. These large browsers all have one pair of procumbent lower incisors, hence the name 'diprotodont'; however, this feature is not unique to the family. Diprotodonty also characterized about a dozen families of Australian marsupials including wombats, koalas, kangaroos, phalangers, palorchestids, wynyardiids and thylacoleonids; all these are sometimes grouped as the order Diprotodonta, though it is possible that the feature may have been acquired independently on several occasions. Diprotodontids have a variable number of upper incisors, one to three pairs. Canines are always lacking and a long diastema separates the incisors from the cheek dentition, which is strongly bilophodont.

Neohelos is a cow-sized diprotodont from the mid Miocene; its abundance in the fossil record suggests it lived in herds, browsing trees and shrubs. Sexual dimorphism is evident with the males being larger than the females. *Zygomaturus* ranged from late Miocene times through almost to the present day. It is especially common in coastal sites and may have favoured forest habitats there. *Zygomaturus* is characterized by bony protuberances on the nasal and frontal bones. *Euryzygoma* from the Pliocene had a large skull mostly filled with sinuses or air spaces; the brain of this bullock-sized animal was no larger than an orange and it has been described as a mental lightweight in the heavyweight class. The skull has a long flared zygomatic process which, combined with the high nasal opening, gave it a very distinctive facial appearance. *Diprotodon* was the largest member of the family. Individuals stood 2 m high at the shoulder and exceeded 3 m in length.

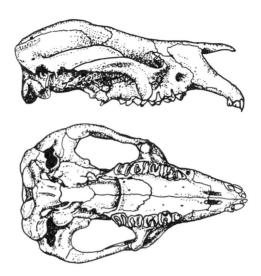

Skull of *Ngapakaldia* (length 25 cm), a mid Miocene palorchestid. Note the high nasal bone, three incisors followed by long diastema and the bilophodont cheek teeth. (After Stirton).

Head and skull (length 50 cm) of the Pliocene diprotodont *Euryzygoma*. Note the very high nasal opening, the long flared process below the ear and the bilophodont teeth.

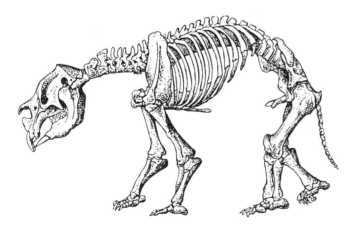

Skeleton of *Diprotodon* (length 3 m). This was the largest of the Pleistocene marsupial herbivores. Heavily built, with short front legs, plantigrade feet and chisel-edged upper incisors. Note also the marsupial bones in the pelvic region.

The skeleton is heavily built, the forelimbs very short and four claws are present on each foot. The plantigrade feet would have given this beast the appearance of a large Alaskan bear. *Diprotodon* skeletons have been recovered in quantity from the dried up Pleistocene clays of Lake Callabonna; perhaps they died there during a drought. The genus is common in Pleistocene deposits in central Australia where the vegetation was probably open woodland and grassland.

Although the kangaroos are the dominant element in the Australian marsupial faunas today, this is a relatively recent event. We can trace their history back to mid Miocene times with tiny primitive mammals that looked like possums and were probably not far from ancestral types. The living musky rat-kangaroo of northern Queensland is a survivor of such a primitive type; the tiny *Hypsiprymnodon* lives in rainforests, is omnivorous, and has a quadrupedal bounding gait rather than the true bipedal hopping of most kangaroos. Macropodoids are probably the sister group to the phalangeroids (possums). There are two families of macropodoids, the potoroids (potoroos and rat-kangaroos) and the macropodids (kangaroos and wallabies). The Macropodidae are a large family by marsupial standards; there are about 50 living species in 13 genera and the fossil list currently doubles these figures. Most are medium- to large-sized and use the bipedal leaping gait. The tibia is greatly elongated, the metatarsus is also elongate and rotation is either reduced or lost. The toes are reduced to three, two or even one digit in the hind foot; the small front limbs retain five clawed digits. The tail is long and strongly developed; it acts as a counter balance when leaping and as a third ground contact when stationary, thus helping to take the weight. Macropodoids have adapted to a very wide range of habitats, paralleling the artiodactyls among the placentals with ecological analogues of deer, tragulids, bovids, hares and lemurs. They have a microflora in the gut to process cellulose vegetation. There are arboreal roos (*Dendrolagus*), burrowing (*Bettongia*), forest, woodland and plains types.

In the late Miocene we see the first of the ancestors of the living stocks of large grazing kangaroos, but it is not until early Pliocene that an explosive radiation of grazing roos occurred. This explosion we can attribute to the increasing aridity which reduced the forests and led to the great expanse of grasslands. The peak of aridity appears to have occurred during the Pleistocene and may coincide with the peak extinction of the marsupial megafaunas, when so many of the large taxa died out. *Prionotemnus* was a medium-sized Pliocene roo with primitive H-pattern molars. *Sthenurus* was a large roo common in the Pliocene and Pleistocene; it had low-crowned molars and very narrow lower incisors. *Procoptodon* was the giant among large kangaroos. It is abundant in Pleistocene deposits and easily recognized by its short face and a single functional hind toe (the fourth). This grazing giant stood over three metres tall.

Restoration scene with elements of the Australian Pleistocene monotreme and marsupial megafauna. Behind: *Neohelos*, cow-sized browsing diprotodont. Front left: *Phascolonus*, giant browsing wombat. Front right: *Zaglossus*, giant montreme anteater.

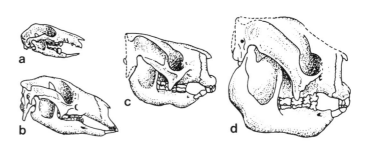

Skulls of four diprotodontoid marsupials; a & b living, c & d fossil:
a) potoroid *Hypsiprymnodon* (length 5 cm)
b) macropodid *Wallabia* (length 10 cm)
c) macropodid *Sthenurus* (length 14 cm)
d) macropodid *Procoptodon* (length 18 cm)

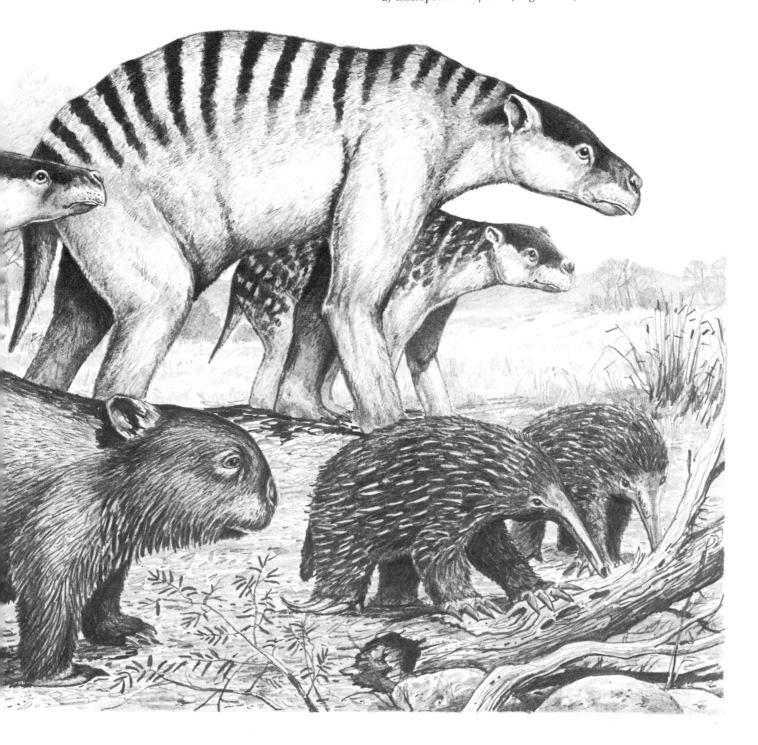

Placental invaders and introductions

The earliest record of a placental mammal in Australia dates from the mid Miocene; a bat is known from both the Ngapakaldi and Riversleigh faunas. No marsupial bat is known so the chiropterans were not in direct competition with the native fauna. It was not until early Pliocene times that the first rodent arrived; its appearance in Queensland suggests it travelled from Papua New Guinea (Irian). Throughout the Caenozoic the Australian tectonic plate has been moving northward towards the Asian plate, narrowing the gap between the two landmasses. During the Plio-Pleistocene, land connections or island chains linked Australia with Papua New Guinea. The Australian mammal fauna extends not only to Papua New Guinea, but eastward to the Solomon Islands and westward to Sulawesi and Timor. The south-east Asian mammal fauna overlaps with this in the west, having distributions in Sulawesi, Moluccas and Timor. The Victorian naturalist Alfred Russell Wallace recognized this watershed between the Australasian and Asian biota and established what is now known as Wallace's Line. There is, however, not just one dividing line but a broad zone which separates the two biota, with a mixing of the faunas and floras in the overlap zone. Elephants, for example, are part of the south-east Asian fauna; they are well known in Pleistocene deposits of Kalimantan (Borneo), Java and Sumatra. They also invaded Sulawesi, Flores and even Timor, but not Australia.

The final changes in the Australian faunas have been brought about by man. The aboriginals introduced the dingo and European man introduced a host of mammals which are rapidly transforming the Australian scene.

Restoration of *Procoptodon* (height 3 m), a Pleistocene giant short-faced grazing kangaroo.

When traps are set for small mammals in the Australian desert, the commonest captives are not the native marsupial dasyurids but *Mus musculus*, the European house mouse. And while collecting the traps, one is quite likely to see a feral camel, a relict of the early exploring days.

Restoration scene with elements of the Australian browsing megafauna. Background: *Diprotodon*. Midground: *Palorchestes*. Foreground: *Zygomaturus*.

CHAPTER 12
Hoofed herbivores

One third of all known mammalian genera are herbivores and more than half of these belong to two orders – the Perissodactyla and the Artiodactyla, meaning odd toed and even toed; one or three toes are usual in perissodactyls (tapirs, rhinos and horses) while two or four toes are usual in artiodactyls (pigs, camels, deer, giraffe, cattle, antelopes, goats and many others). This vast array of herbivores is formally classified into 12 families and there are double that number of extinct families. They range from rabbit-size to the bison, with many large extinct forms; one rhinoceros was the largest known land mammal. Their origins are probably in the Palaeocene and the fossil record for both orders is good. Perissodactyls are dominant in the early Tertiary, with the artiodactyls gradually taking over until today there are only six genera of perissodactyls but around 80 artiodactyl genera. The families appear mostly to have arisen on northern continents and later spread southward into Africa and South America; none ever reached Australia. Most early Tertiary stocks are rooters and browsers, many dying out to be replaced by colonizers of the savannahs and open grassland habitats of the later Tertiary. Diversification into a very wide range of habitats was achieved by major anatomical and physiological adaptations. Limbs tended to elongate, particularly the feet, many became digitigrade and even unguligrade; the number of hoofed toes was reduced to three, two or even one. These modifications, combined with alterations in the ankle joints and fusion of the paired limb bones, enabled the herbivores to exploit fully a vast range of terrains, move efficiently over long migrations and rapidly escape predators.

The dentitions of early genera were almost complete; in later stocks upper incisors are often lost, the lower incisors biting against a hard pad. There is also a tendency for the canines and anterior premolars to be reduced, leaving a diastema in front of the battery of grinding teeth, which become high crowned, develop complex cusp patterns and add cementum to the crown. The dental changes, which usually involve elongation of the face, reach their acme among the horses.

The alimentary canal or gut of herbivores is much more complex than that of insectivores or carnivores. Digestion of cellulose is made possible only by the presence of micro-organisms in the gut. The stomachs of many ruminant artiodactyls are subdivided into a series of chambers allowing specialized functions associated with ruminating or redigestion of food before it is passed

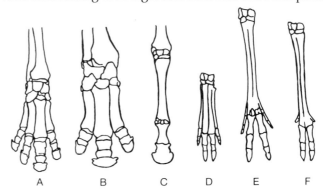

Forefoot of both odd (A–C) and even (D–F) toed ungulates to show the reduction of the digits in each. A Tapir (4 toes); B Rhinoceros (3 toes); C Horse (1 toe); D *Leptomeryx* (2 toes and 2 reduced toes); E *Blastomeryx* (2 toes and 2 vestigial toes); F *Merycodus* (2 toes).

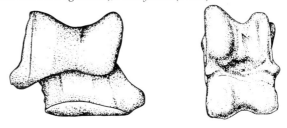

The astragalus (ankle bone) of a perissodactyl (left) and an artiodactyl (right). The double pulley (top and bottom) is characteristic of the artiodactyl astragalus.

The relative abundance of odd (perissodactyl) and even (artiodactyl) toed ungulates through Caenozoic time: beginning 60 million years ago faunas were dominated by perissodactyls, but as time progressed these were gradually replaced by artiodactyls so that today they make up 90 per cent of ungulate faunas.

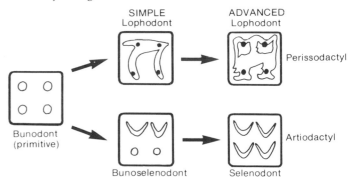

Diagrammatic sequence of evolutionary changes from a simple four cusped bunodont ancestral form of molar tooth into typical perissodactyl and artiodactyl patterns.

to the intestines. Thus ruminants make very efficient use of the vegetation they consume; almost nothing is wasted.

Several lineages independently develop skull outgrowths, primarily as defence mechanisms. Horns develop in both sexes in bovids (cattle and antelopes for example). A pair of conical bony outgrowths develops on the frontal skull bones; in life these are covered with a horny sheath. The horns may be straight, curved or spiral but never branch and are permanent, not shed annually. By contrast, antlers develop in cervids (deer), usually only in the male though they are present in both sexes of reindeer. They are paired, branching, bony outgrowths of the frontal skull bones, lacking a horny sheath; they grow and are shed annually. Ossicones in giraffes are bony outgrowths with a permanent covering of skin. In rhinoceroses the nasal 'horn' is composed of compact hair and so does not fossilize. Several extinct stocks (e.g. titanotheres, uintatheres and protoceratids) had bony skull outgrowths, usually referred to loosely as

'horns'; they were never shed, some were branched, their number and head site varied, they were usually larger in the male, but whether they were covered in skin or horn is unknown.

All early members of these two orders were small and hornless. When they reached a weight of about 20 kg, horns began to develop and sexual dimorphism appears. It is probable that early taxa were forest dwellers; as they moved out into woodlands, savannah and grasslands many became territorial and in consequence needed some defence for the herd.

Domestication of wild animals is one of the greatest achievements of mankind. In western Asia by the seventh millenium BC (about 9000 years ago) goats and sheep were being domesticated, to be followed by cattle, pigs, horses and camels. Over the succeeding millenia man was to alter these herbivores by selective breeding for his own ends. Their importance in the economy of man is perhaps most clearly reflected in the dramatic rise in human populations from around 100 million people in the Neolithic (about 10 000 years ago) when domestication began, to a staggering 4500 million today. Domestication of livestock must rank with crop cultivation and medicine as one of the three major factors in this explosion.

The basic pattern of the upper and lower molar teeth as seen in tapirs and rhinoceroses.

Early and primitive ungulate teeth are all brachydont (low crowned, left) and most advanced forms are hypsodont (high crowned).

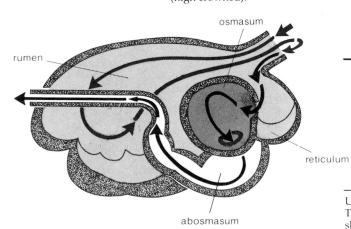

Diagrammatic representation of a ruminant stomach. The black arrows follow the food from the mouth into the first two stomach chambers – the rumen and reticulum. The rumen is a fermenting vat where micro-organisms break down the cellulose plant material and produce amino acids and proteins. After regurgitation and chewing of the cud, the food follows the brown arrows to the third chamber – the osmasum – where excess water is removed, and then to the fourth chamber – the abosmasum – where conventional digestion proceeds before the residue passes into the intestine.

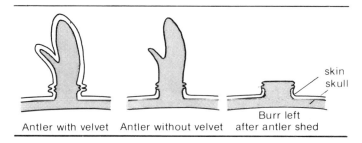

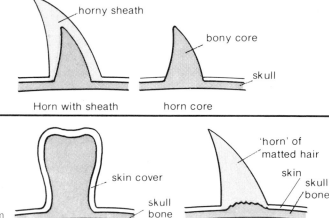

Ungulate skull outgrowths.
Top: antlers grow annually on male deer (rarely on female), and are shed each spring. They are branched bony structures with a skin covering during growth. When shed a burr is left on the head.
Middle: horns grow on male and female bovids (cattle, sheep, antelopes). They have a bony core and a horny sheath. They do not branch and are not shed.
Bottom, left: ossicones develop in giraffes; they may or may not branch, they have a skin covering during life and are not shed.
Bottom, right: rhinoceroses may have one or two 'horns'; these are composed of matted hair, are larger in males and continue to grow throughout life.

Odd toed ungulates

Also known as the Perissodactyla or Mesaxonia, in these animals the axis of weight bearing passes through the middle or third digit. Most members are three toed, but the later horses have eliminated the lateral digits to become one toed. There are five major stocks – tapirs, rhinoceroses, horses, and the extinct brontotheres and chalicotheres. Of the 14 families only three survive today (tapirs, rhinoceroses and horses). The order had its peak in the Eocene but by the end of Oligocene times ten of the 14 families were extinct. The perissodactyls dominated the medium- and large-sized herbivores of the early Tertiary. They have a good fossil record in North America and Eurasia, with later members penetrating southward to Africa and South America.

ODD TOED HERBIVORES

	Examples
Order PERISSODACTYLA	
Suborder CERATOMORPHA	
Superfamily Tapiroidea	
Family † Helaletidae	† *Heptodon*
Tapiridae	*Tapirus*
† Hyrachyidae	† *Hyrachyus*
Superfamily Rhinocerotoidea	
Family † Hyracodontidae	† *Hyracodon*
	† *Forstercooperia*
	† *Indricotherium*
† Amynodontidae	† *Metamynodon*
	† *Cadurcodon*
Rhinocerotidae	† *Caenopus*
	† *Teleoceras*
	† *Aceratherium*
	† *Coelodonta*
	† *Elasmotherium*
Suborder ANCYLOPODA	
Superfamily Chalicotheroidea	
Family † Eomoropidae	† *Litolophus*
† Chalicotheriidae	† *Moropus*
	† *Chalicotherium*
Suborder HIPPOMORPHA	
Superfamily Equoidea	
Family † Palaeotheriidae	† *Palaeotherium*
Equidae	† *Hyracotherium*
	† *Mesotherium*
	† *Miohippus*
	† *Anchitherium*
	† *Hypohippus*
	† *Parahippus*
	† *Merychippus*
	† *Hipparion*
	† *Pliohippus*
	† *Hippidion*
	Equus
Superfamily Brontotheroidea	
Family † Brontotheriidae	† *Eotitanops*
(Titanotheres)	† *Palaeosyops*
	† *Brontotherium*

Tapirs

The first tapirs appear in the early Eocene at the same time as the first horses and chalicotheres, with brontotheres and rhinoceroses present before the end of the Eocene. This implies an origin and early differentiation of the order back in the Palaeocene. *Heptodon* is one of the earliest tapirs and from the early Eocene of Wyoming are known skulls and a partial skeleton. The living tapir, despite some 55 million years of 'evolution', still bears a remarkable similarity to its ancestor. *Tapirus* is about twice the size of *Heptodon*; it retains four toes on the fore foot and three on the hind foot; the toes are hoofed with a supporting pad. The ulna and fibula are complete and unfused. All these limb characters of the living tapir are seen in its Eocene ancestor *Heptodon*. The skulls are also similar in structure, except that there was no proboscis in *Heptodon*. The dentition of tapirs is complete though a small diastema exists; the upper canine is reduced and the lateral incisor has become caniniform. In *Tapirus* all but the first premolar have become molariform, but the molars in both *Heptodon* and *Tapirus* are closely similar. The basic tapiroid (and rhinocerotoid) pattern is a three loph upper molar, the lophs or ridges arranged as in the Greek letter pi π; the lower molars have two transverse lophids.

Tapirs are tropical browsing mammals, living in forests and woodlands near water; they survive today in Central and northern South America and in south-east Asia, remnants of a once widespread distribution. Tapirs persisted in the warmer parts of North America, Europe and Asia until late Pleistocene times.

Left: skull of *Heptodon* (length 20 cm), a primitive tapir from the early Eocene, and right skull of *Tapirus* (length 36 cm), a living tapir. The nasal bones (above the eyes) on the living tapir are indicative of a well developed proboscis. (After Radinsky)

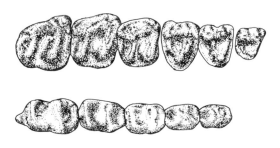

Top, the upper, and above, the lower cheek dentition of *Heptodon*; although the earliest known tapir, it shows characteristic tapir features; the upper molars have a π pattern and each of the lower molars has two transverse lophs. (After Radinsky).

Tapirs are arranged in about five families, with the Hyrachyidae forming a transitional family between the tapirs and the rhinoceroses. *Hyrachyus* is found abundantly in mid Eocene times in North America and Europe. It is very similar to *Heptodon* but larger and its teeth show some rhinoceros features; on the upper molars the metaloph is long and flat, the parastyle reduced and on the lower molars the lophids are high.

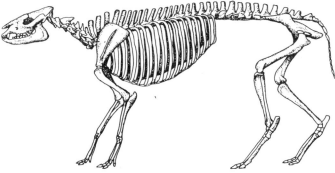

Skeleton of *Heptodon* (overall length 1 m), a tapir from the early Eocene of Wyoming. It was not dissimilar from a half grown tapir, but lacked the proboscis. (After Radinsky)

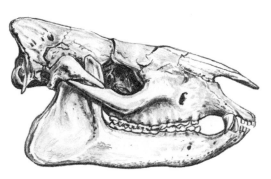

Skull of *Heptodon* with complete dentition and a well-developed diastema between the canine and anterior premolars. (After Radinsky.)

Restoration of *Hyrachus*, a mid Eocene tapir from North America. This animal was midway between a tapir and a primitive rhinoceros.

Restoration of *Heptodon*, a 55-million-year-old tapir from Wyoming.

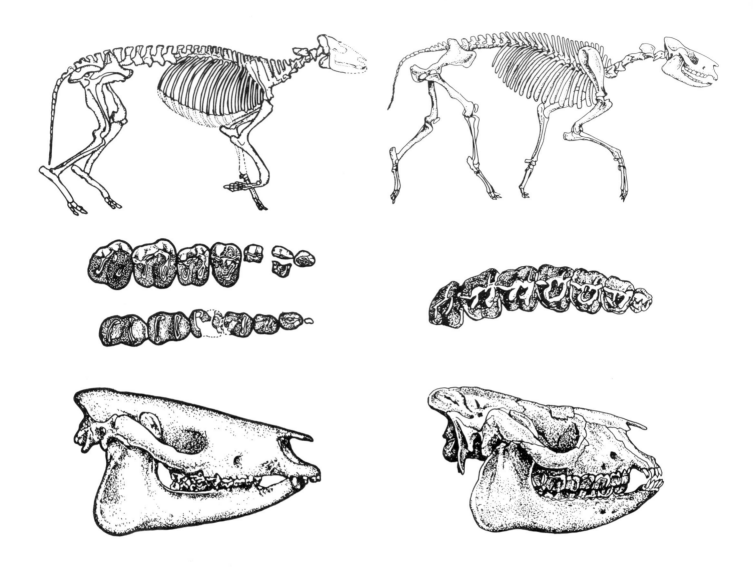

Skeleton (overall length 1.6 m), cheek dentition and skull (length 30 cm) of *Hyrachyus*. The nasal incision is larger than in *Heptodon* and the teeth show traces of rhinocerotoid characters. (After Wood).

Skeleton (length 1.5 m), dentition and skull (length 27 cm) of *Hyracodon*, a running rhinoceros from the Oligocene of North America. This animal was lightly built like a horse, had three toes on each foot and the posterior premolars are molarized. (After Scott).

Rhinoceroses

From a tapiroid close to *Hyrachyus* arose two lineages of rhinocerotoids – the hyracodontids and amynodontids – both stocks abundant in the Eocene and Oligocene of North America and Asia. The hyracodontids comprise about a dozen genera; most of these were medium-sized like *Hyracodon*. Their elongate almost slender limbs and light build parallels that of horses, and like them they were cursorial. Their posterior dentition with progressive molarization of the premolars and increase in crown height was little changed from that of tapirs, though there were many variations in the anterior teeth. The incisors were primitively spatulate and equal-sized and the canines of moderate size. However, in Oligocene times in Asia there arose from a hyracodont close to *Forstercooperia* a series of gigantic hornless rhinoceroses, which may be grouped in the subfamily Indricotherinae. They ranged from Caucasia, through

Restoration of *Hyracodon*, a hornless three toed running rhinoceros of the Oligocene period.

Central Asia and Baluchistan to China. *Indricotherium* (= *Baluchitherium*) is the largest land mammal that ever lived. This rhinoceros stood 5·4 m at the shoulder; with its long neck and 1·3 metre-long skull it could reach to browse on vegetation over eight metres above the ground. With a probable weight of around 30 tonnes it was four and a half times as heavy as the heaviest recorded elephant (6·6 tonnes) and nearly twice the estimated weight of the heaviest mammoth.

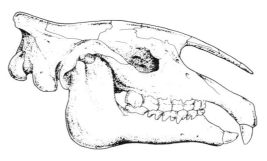

Skull (after Granger & Gregory), size comparison and restoration of the 35-million-year-old rhinoceros from Central Asia *Indricotherium* (= *Baluchitherium*), the largest land mammal that ever lived.

The second Palaeogene lineage of rhinoceroses comprised the amynodontids, a compact group of about ten genera with large heavy bodies, short stocky limbs and short faces with prominent canine tusks. Among their number was *Metamynodon* which had prehensile lips and *Cadurcodon* with a short tapir-like proboscis. This, combined with the very hippopotamus-like skeleton found among members of this family, strongly suggests a semi-aquatic habitat.

The great radiation of Palaeogene rhinoceroses was all but over by the close of the Oligocene, leaving only a few stragglers to see the earliest Miocene times in Asia. However, there arose from among the hyracodontids during the Oligocene a lineage with distinct specializations of the incisor teeth, namely the development of chisel-shaped I^1 and procumbent hypertrophied lanceolate or tusk-like I_2; these were to form the basis of a second great radiation of rhinoceroses in the Neogene – the Rhinocerotidae. They comprise about 50 genera, including four present-day survivors; they were abundant in North America, Eurasia and Africa during the Miocene and the Plio-Pleistocene, and adapted to a wide range of habitats, climates and diets. Many were browsers as were earlier rhinoceroses, but some became specialized grazers. Some had thick coats of long hair, many had 'horns' which were composed of matted hair and do not fossilize, but a bony protuberance on the skull is indicative.

Caenopus is representative of the early rhinocerotids in the early Oligocene of North America. It was very tapir-like with short stout limbs, four-toed front feet and three-toed hind feet; the last premolar was molarizing and the cheek teeth were low crowned. *Teleoceras* was a hippopotamus-like rhinoceros from the Miocene of North America. It had a bony rugosity on the nasal bone implying the presence of a horn; though *Teleoceras* was similar in size to living rhinoceroses, it had a brain that was twice as large. *Aceratherium* and its allies were medium-sized rhinoceroses abundant in the Miocene of Eurasia and Africa. They were hornless, had reduced incisor teeth and had subhypsodont molars suggesting a mixed diet of foliage and grasses.

Restoration of the Ice Age woolly rhinoceros, *Coelodonta antiquitatis*, a large furry rhino with two nasal horns that roamed the tundra of Palaearctica during the late Pleistocene.

Opposite: restoration of *Elasmotherium*, a very large rhinoceros from the mid Pleistocene of the steppes of southern Siberia; the single horn on the forehead could reach 2 metres in length.

The five species of living rhinoceroses are to be found in tropical Africa, India, Sumatra and Java. The Great Indian rhinoceros (*Rhinoceros*) and the White African rhinoceros (*Ceratotherium*) may each approach 4 tonnes in weight, the largest land mammals today after the elephants. They live in woodland savannah, grazing and browsing and each has two nasal horns. Other rhinoceroses are rather smaller, with one or two horns and are forest browsers. During the Pleistocene there was a large diversity of rhinoceroses in different parts of the world. The woolly rhinoceros (*Coelodonta*) was widespread from Britain to eastern Siberia, but never crossed Beringia into America. It is clearly depicted in the cave paintings of Palaeolithic man with its two large horns and long shaggy coat. The nasal region was supported by an ossified medial septum; the incisors were absent and the cheek teeth very high crowned. Like the woolly mammoth, the woolly rhinoceros was highly adapted to grazing the temperate and tundra grasslands. Another Pleistocene rhinoceros was *Elasmotherium* from the steppes of southern Russia; this was a very large rhino with a skull 75 cm long. Instead of a nasal horn, it had an enormous 2-metre-long horn originating on the forehead, which has given rise to its alternative name, the giant unicorn. The incisors were lacking and it would have used its lips to pluck grasses. The cheek teeth have very complex enamel patterns, cementum and extremely high crowns; it was the most specialized grazer evolved among the rhinoceroses.

Cave painting in red ochre of a woolly rhinoceros by late palaeolithic man from Font-de-Gaume, southern France.

Above; restoration, and below, skeleton of *Caenopus* (length 2.5 m) and skull (length 50 cm), an early true rhinocerotid from the Oligocene of North America. (After Osborn & Scott).

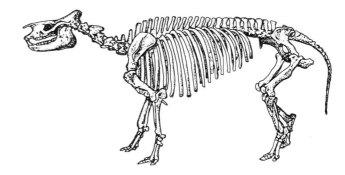

Restoration of *Teleoceras* from the Miocene of Kansas, a hippopotamus-like rhinoceros with a small nasal horn.

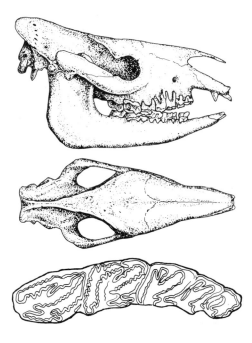

Restoration of *Cadurcodon*, an amynodont rhinoceros with elongate proboscis, from the Oligocene of Mongolia.

Skull of *Elasmotherium* (length 75 cm) and cheek dentition (length 22 cm). The tooth enamel is characteristically heavily crinkled. (After Brandt).

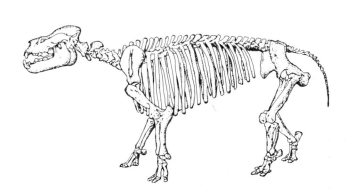

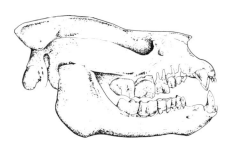

Restoration, skull (length 56 cm) and skeleton (length 4 m) of *Metamynodon*, a large semi-aquatic hippo-like rhinoceros from the Oligocene of North America. Note the large canine tusks. (After Scott & Osborn).

Chalicotheres

Periodically come reports from the Kakamega forests in Kenya of sightings of the Nandi bear. The beast is described as having a gorilla-like stance with forelimbs longer than the hind, with clawed feet like a bear and with a horse-like face. Could the beast be a survivor of the chalicothere, thought to have become extinct in East Africa during the Pleistocene? The description above would fit with the skeletal remains of these

Restoration of *Chalicotherium grande*. This strange browser had very short hind legs and walked on the knuckles of its forefeet which had long inwardly turned claws.

extraordinary animals, which though never abundant in the fossil record, are known from Eocene times onward and have been found in North America, Eurasia and Africa. *Litolophus* from the Eocene of Mongolia was about sheep-sized. *Moropus* from the Miocene of America was horse-sized with an elongate horse-like face and some had domed skulls; the cheek teeth retained simple premolars and the molars were bunolophodont. It was probably a bipedal browser, using its long forearms and curved claws to hook down leafy branches in the wooded savannahs in which it lived and it may additionally have used its claws to uproot tubers for water in the dry season. *Chalicotherium* itself is known from the Miocene of Eurasia and Africa; it was a forest browser, which may account for its rarity in the fossil record. While the skull and dentition are typical of perissodactyls and closely similar in many ways to titanotheres (see below), the limbs with their curved claws are very un-ungulate like, and reflect a specialized adaptation to bipedal browsing.

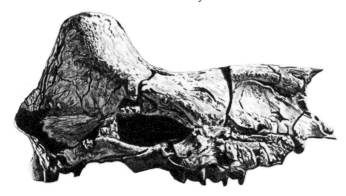

Skull (length 30 cm) of a chalicothere from the Miocene of Wyoming with domed roof; probably used by males in sparring contests. (After Munthe & Coombs).

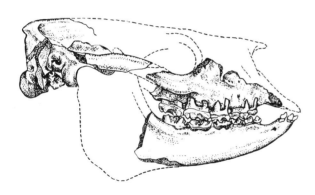

Skull of *Chalicotherium grande*, a large chalicothere from the European Miocene.

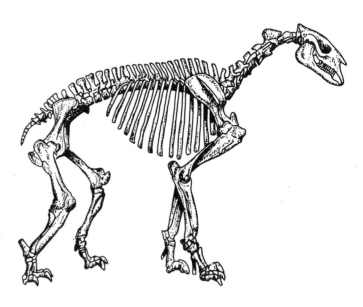

Above, skeleton (after Gregory), and below, restoration of *Moropus*, a 3-metre-long chalicothere from the Miocene of North America. Note the sloping hindquarters, short hind legs and peculiar clawed feet that were probably used to pull down branches to reach the foliage.

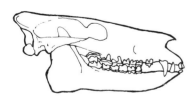

Top, skull (length 30 cm), and above, dentition of *Litolophus*, a late Eocene eomoropid from Mongolia; this animal was a sheep-sized relative of the chalicotheres. (After Colbert & Radinsky).

Horses

The evolution of horses is often quoted as a prime example of evolutionary gradualism; the progressive changes over 55 million years from a small lamb-sized eohippus to the modern horse involved increase in size and height, and in the complexity of the enamel pattern on the cheek teeth, in length of legs and in reduction of the toes from three to one. All these changes are viewed as leading to a faster and more efficient grazing mammal. In terms of macroevolution such trends can be unambiguously discerned, but the details reveal a much more complex picture. The identification of genera requires not just dentitions but skulls, in which the presence or absence of preorbital fossae is regarded as significant, although as none exist on living horses their purpose is unknown; different genera appear sometimes to have very similar dentitions. The evolutionary patterns involve numerous multiple migratory events and parallel development of characters. Our current knowledge is not sufficient to resolve the details of horse evolution, but the broad outlines can be sketched with reasonable assurance.

Hyracotherium (= *Eohippus*) is the first known horse, found in early Eocene deposits of North America, western Europe and eastern Asia. It can be derived from a phenacodont condylarth, from which it differs little in dental characters having a complete dentition with six-cusped low-crowned squarish upper molars. However, the brain of *Hyracotherium* is vastly different from that of a condylarth; it is much more advanced with marked convolution of the cerebral cortex and complexity of the cerebellum. These early horses were alert and intelligent little animals with a short face, four hoofed toes on the forefoot and three on the hind and a cushioned pad to help bear the weight. In western Europe the hyracotherines overlapped in their savannah-woodland territories with their forest-dwelling near relatives the palaeotheriids; these browsers were of a tapir build and some were as large as a rhinoceros, with short stout limbs. The Eocene *Palaeotherium* had some very advanced horse characteristics; the premolars were molarized, the molars were lophodont, subhypsodont and bore cementum – all features which do not appear in true equid horses until the Miocene. Yet despite these

Tibia Fibula	Hind foot	Ulna Radius	Side	Front foot With soft tissues	An

Restoration of *Palaeotherium*, a tapir-like horse relative from the Eocene of Europe.

Evolution in a series of six horse genera from the ancestral *Hyracotherium* in the Eocene to the living *Equus*. The sequence of morphologoical changes through time in these horses is not a direct phylogenetic lineage, for throughout horse evolution there have been divergent stocks and adaptive radiations. The changes in the limbs and the dentition are each marked by a major transformation. In the dentition it is the change from browsing horses with brachydont or low crowned teeth to grazing horses with hypsodont or high crowned teeth, and this occurred in mid Miocene times. The changes from three toed to fully functional one toed horses occurred at the end of Miocene times; horses that had adapted to eating grasses were then able to run more efficiently over the grasslands and escape predators.

		Skull	Brain	Upper molar Crown	Upper molar Side	
Recent and Pleistocene	*Equus*					Grazers
Pliocene	*Pliohippus*					Grazers
Miocene	*Merychippus*					Grazers
Miocene	*Parahippus*					Browsers
Oligocene	*Mesohippus*					Browsers
Eocene	*Hyracotherium*					Browsers

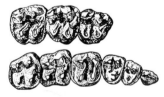

Upper cheek dentition of *Hyracotherium* (bottom) and *Orohippus*, an early and a late Eocene horse. In *Orohippus* the six-cusped molars have become more lophodont. (After Cope and Granger).

Front foot, skull and upper cheek dentition of *Palaeotherium* from the Eocene of France. Palaeotheres were closely related to horses, had a short proboscis like a tapir, but retained primitive unspecialized feet. (After von Koenigswald & Stehlin).

Restoration of *Hyracotherium*, the earliest known horse, from the late Palaeocene and early Eocene of both North America and Europe.

advanced features, the palaeotheres became extinct during the early Oligocene; although the true horses had themselves also by then died out in Europe and Asia, they survived in North America and from there they continued to evolve.

The Oligocene equids are *Mesohippus* and *Miohippus*. They were about the size of sheep, the first true three-toed horses, with the middle digit rather larger than the laterals. The snout was elongating, the premolars show signs of molarization and the battery of six grinding cheek teeth have lophs and lophids. These horses were adapted to the more arid savannahs of the Oligocene in North America. From them arose a sideline in the early Miocene with *Anchitherium*, a very successful horse which migrated through Europe and Asia. Toward the close of Miocene times the large forest-dwelling *Hypohippus* migrated into China.

From the Oligocene anchitheres arose in the early Miocene *Parahippus*, a structural and temporal intermediate step to the equine *Merychippus* of the mid Miocene. *Merychippus* marks a major step in horse evolution – the first grazing horse. It had truly hypsodont cheek teeth, the high crowns extending below the gum line into the sockets; the teeth were also elaborately lophed and had cementum. It was thus able to exploit the increasingly arid prairie grasslands which prove very abrasive on the dentition. Its long limbs with fused ulna/radius and tibia/fibula enabled it to gallop without danger of twisting its wrist and ankle joints, and so escape predators more readily. *Merychippus* has a long face and was not dissimilar to a living horse, though still three-toed; being smaller it matured early at 3 years. From *Merychippus* (sensu lato) evolved all later horse lineages, living and extinct. Of the extinct sidelines the most important were the hipparionines, with perhaps as many as six separate lineages. The hipparion horses are characterized by very complex enamel patterns on the cheek teeth which makes them readily recognizable.

Hipparions appear to have originated about 15 million years ago in North America and several waves of them successively invaded the Old World. *Hipparion* (sensu lato) is known from Britain to China, from Russia to South Africa, and in some areas survived into the late Pleistocene. The late hipparions had very reduced lateral toes so that they were almost one-toed horses. They very fully exploited the grasslands and savannahs of the Old World; there appear to have been temporal, geographical and ecological variants. Many species have been named but their detailed evolution is not clearly understood.

Restoration of *Parahippus* from the early Miocene of North America, the last browsing three toed horse.

Restoration of *Mesohippus*, an Oligocene browsing horse, bigger than *Hyracotherium* and with three toes on each foot.

Restoration of *Anchitherium*, a side line of three toed horses that colonized Eurasia from America during the Miocene and then became extinct without leaving any descendants.

In late Miocene times *Merychippus* was replaced by *Pliohippus*, the first one-toed horse, and it in turn gave rise to *Equus* during the Pliocene, and in Central America to *Hippidion* which invaded South America. However, it was *Equus* that was to become like *Hipparion*, a widely established horse around the world, exploiting the arid steppes, prairies and savannahs of Europe, Asia and Africa. Today the genus comprises zebras, onagers, asses, wild and domestic horses. With *Equus* we see the product of 55 million years of adaptation to ever increasing aridity, to more efficient feeding on grasses, to larger overall size and more rapid locomotion. North America has been the centre of evolution of horses throughout their history. During the Pleistocene both New World continents abounded in them and then, some 8000 years ago, the last wild horses in the Americas became extinct, though fortunately they survived in the Old World. The cause of their sudden demise is unknown; a possible but untestable explanation is that a virulent disease wiped them out; being herd, migratory and slow breeding mammals, a myxomatosis-like disease could effect such a disaster. Man first domesticated the wild horses of Asia around 1500 BC; since that time he has used artificial selection to breed varieties ranging from the 0·4 metre-high Falabella horse to the 2 metre-high Shire horse.

Titanotheres

Titanotheres (= brontotheres) have much in common with uintatheres; both were medium- to very large-sized herbivores of the early Tertiary in North America and eastern Asia, both fed on soft folivorous vegetation of the forests, and both had stocks that developed horns. They overlapped in time but with successive acmes; and both became extinct soon after reaching their acme; the uintatheres peaked in the late Eocene and the titanotheres in the early Oligocene, to be ecologically succeeded in Asia by the giant indricothere rhinoceroses in the later Oligocene. The details of the dentitions, skull and horns show the titanotheres and uintatheres to be quite separate in origin and the similarity represents convergent evolution.

Eotitanops is an early Eocene eohippus-like titanothere; it shares with horses a common phenacodont ancestry.

Merychippus from the late Miocene of North America, a descendant of *Parahippus* with high crowned teeth capable of grazing on prairie grasses.

Restoration of *Eotitanops*, a primitive titanothere from the early Eocene, little different from its contemporary, the horse *Eohippus*.

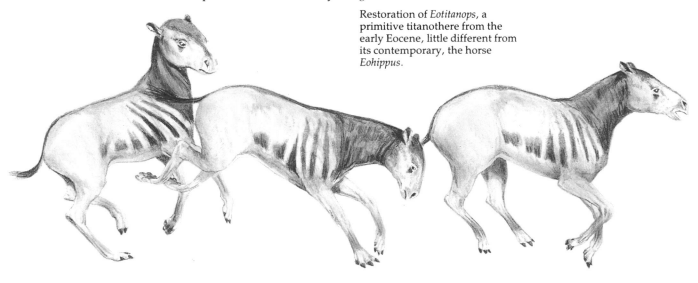

Pliohippus, the first one toed grazing horse. It appeared during late Miocene times, became abundant during the Pliocene and gave rise in Pleistocene times to the living genus *Equus*, the horses and zebras.

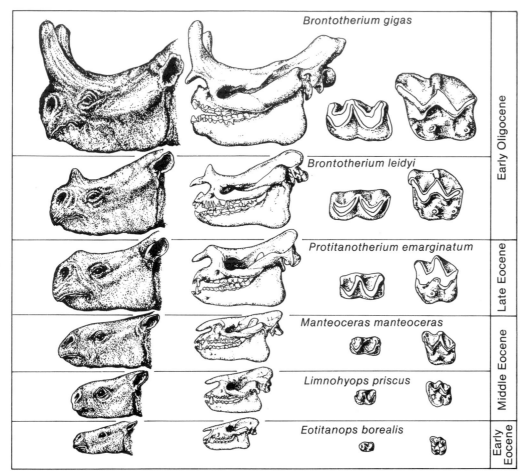

Brontotherium gigas

Brontotherium leidyi

Protitanotherium emarginatum

Manteoceras manteoceras

Limnohyops priscus

Eotitanops borealis

Early Oligocene

Late Eocene

Middle Eocene

Early Eocene

A series of six North American titanothere species ranging from the early Eocene to the early Oligocene. They display an increase in skull size and development of a large nasal horn. The upper and lower cheek teeth increase in size but the pattern remains constant throughout.

Restoration and skull (length 40 cm, after Scott & Osborn) of *Palaeosyops* from the mid Eocene of North America. *Palaeosyops* was larger than *Eotitanops*, and no nasal outgrowth has yet appeared.

From *Eotitanops* evolved a number of lineages of medium-sized hornless and larger forest titanotheres; some such as *Palaeosyops* developed robust canines, again a feature seen in uintatheres. Others grew very large; *Brontotherium* in the early Oligocene stood 2·5 m at the shoulder, larger than any living rhinoceros. The long skull is high posteriorly, dipped over the very small brain case and anteriorly has a pair of nasal horns, which are bigger in males than females. The bony horns were rugose and probably covered in skin as in giraffes. The teeth remained relatively simple bunolophodont in pattern, capable only of dealing with soft vegetation. The body was almost elephantine in build with graviportal limbs. Titanotheres were abundant in the early Oligocene of North America and eastern Asia and even penetrated eastern Europe.

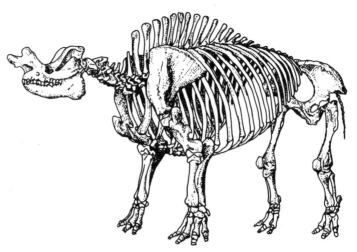

The skeleton of *Brontops* (shoulder height 2·5 m) a titanothere from the early Oligocene of USA. (After Scott & Osborn).

Restoration of *Brontotherium*, a gigantic Oligocene browsing ungulate from North America; these were the rhinoceroses of their day.

EVEN TOED HERBIVORES

		Example
Order ARTIODACTYLA		
Suborder SUINA		
Family	† Dichobunidae	† *Diacodexis*
	† Entelodontidae	† *Dinohyus*
		† *Entelodon*
	† Anthracotheriidae	† *Merycopotamus*
		† *Anthracotherium*
		† *Bothriodon*
	Tayassuidae	† *Perchoerus*
	Suidae	† *Palaeochoerus*
		† *Microstonyx*
		† *Listriodon*
		† *Kubanochoerus*
		† *Tetraconodon*
		Phacochoerus
		† *Metridochoerus*
Suborder TYLOPODA		
Family	† Cainotheriidae	† *Cainotherium*
	† Xiphodontidae	† *Xiphodon*
	† Merycoidodontidae	† *Merycoidodon*
		† *Leptauchenia*
		† *Brachycrus*
	† Agriochoeridae	† *Agriochoerus*
	Camelidae	† *Protylopus*
		† *Poebrotherium*
		† *Protolabis*
		† *Stenomylus*
		† *Oxydactylus*
		† *Aepycamelus*
		† *Titanotylopus*
		† *Camelops*
	† Protoceratidae	† *Protoceras*
		† *Kyptoceras*
		† *Synthetoceras*
Infraorder TRAGULINA		† *Hypertragulus*
		† *Archaeomeryx*
		† *Blastomeryx*
		† *Dorcatherium*
Suborder PECORA		
Superfamily	Giraffoidea	† *Climacoceras*
		† *Palaeotragus*
		† *Samotherium*
		† *Giraffokeryx*
		† *Sivatherium*
	Cervoidea	† *Euprox*
		Cervus
		† *Eucladoceros*
		† *Megaloceros*
Superfamily	*incertae sedis*	† *Hoplitomeryx*
Superfamily	Bovoidea	
Family	Antilocapridae	† *Merycodus*
		† *Hayoceros*
		† *Ilingoceros*
		† *Hexameryx*
		† *Osbornoceros*
	Bovidae	† *Eotragus*
		† *Tsaidamotherium*
		† *Pelorovis*
		Bison
		Bos

† = Extinct

Even toed ungulates

In the Paraxonia or artiodactyls the foot axis is between the third and fourth digits, which are equal in size and whose metapodial elements often fuse to form the cannon bone. The first digit is lost in most but the very early forms, but the second and fifth digits are present in almost all early Tertiary taxa, though variously subequal or reduced; these lateral digits are lost or functionally redundant in all advanced forms. The astragalus or ankle bone has a unique double pulley structure in all members of the order; this imparts great flexibility to the foot and greatly aids springing actions.

The dentitions of artiodactyls differ greatly from those of perissodactyls. Primitive stocks such as pigs have fairly complete dentitions, with often enlarged canine tusks. In later stocks the upper incisors are reduced or lost, a hard pad taking their place as in cattle. Behind the diastema the premolars become partially molarized. The molars are in early stocks brachydont and bunodont, later bunoselenodont and finally both totally selenodont and high crowned. Artiodactyl teeth never attain the complexity seen in horses; however, they have developed very complex stomachs to digest vegetation. In Suina (with pigs and hippos) the stomach has two or three chambers in living forms and they are non-ruminating. Tylopoda (with camels and llamas) have three-chambered stomachs and ruminate. Pecora (with cattle and deer) have four stomach chambers. In the first two (rumen and reticulum) cellulose fibres are fermented by micro-organisms and reduced to the cud. This is then regurgitated and chewed; on reswallowing, the cud enters the third and fourth chambers where digestion continues. This elaborate rumination cycle enables an animal to feed rapidly and digest at leisure; further the food is very thoroughly processed and almost all its nutritive value extracted.

The origin of artiodactyls in the Palaeocene is still obscure. The first great radiation in early Eocene gave rise to many pig-like stocks, rooters and browsers of the forests and woodlands. The second radiation in late Eocene and early Oligocene gave rise to the early ruminants, which in Miocene times very successfully diversified to exploit the savannahs and grasslands where they still remain today the dominant herbivores. Artiodactyls are of prime importance to human economy, constituting the vast majority of our domestic livestock – cattle, sheep, goats, pigs, camels, reindeer and many others.

Dichobunids

From the early Eocene of Pakistan comes a species of *Diacodexis*, the earliest and the most primitive known artiodactyl. Members of the family Dichobunidae, these creatures were no larger than a rabbit. Their ankle bones reveal the diagnostic double pulley in the astragalus; the feet were four-toed with small hoofs. The dentition was complete and the molars surprisingly primitive, being triangular with three bunodont cusps – a persistence of Palaeocene characters found in condylarths such as hyopsodonts. The presence of about six genera of dichobunids in Asia, western Europe and North America in the early Eocene, together with another four genera in other families, points to an origin for the order well back in Palaeocene times. By the close of Eocene times about 20 artiodactyl families are recorded and on all northern continents. This demonstrates the rapid and successful early diversification of the first artiodactyl radiation, but of all these early stocks only the camelids survive today. Most of the extinct families are as yet little known, but we may examine two for which there is a good fossil record – the entelodontids and the anthracotheriids.

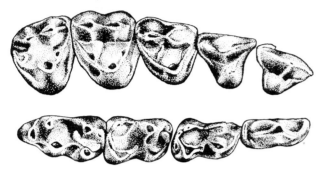

Upper and lower dentition of *Diacodexis* (length 17·5 mm), the most primitive known artiodactyl, from the early Eocene of Pakistan. (After Gingerich).

Skull, skeleton (length 3 m) and restoration of *Dinohyus*, a gigantic entelodont from the Early Miocene of North America, which fed on ground vegetation and probably also scavenged. (After Peterson).

Entelodontids

Entelodonts were large pig-like animals which lived in forests of the northern continents, mainly during the Oligocene. One of the last and the largest was *Dinohyus* in the early Miocene of North America. It was almost hippo-sized – taller but less massive – and with a metre-long skull. Most genera had, as the family name implies, a complete dentition. The incisors were often stout and blunt and the canines robust, suggesting they could have been omnivorous feeders like pigs. The molars

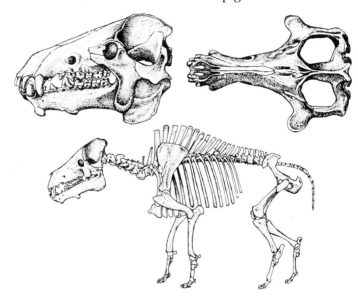

remained simple five-cusped bunodont teeth with thick enamel. The postorbital bar was complete (rare in early artiodactyls) and the small brain had large olfactory lobes suggesting a good sense of smell, perhaps for sniffing out underground tubers. The neck and limbs were short, the radius and ulna fused, although no cannon bone formed, and the lateral toes were vestigial. The most remarkable feature of entelodonts was the extraordinary development of a heavy zygomatic arch on the skull with a large ventral flange, while on the mandible there developed two pairs of prominent bony tubercles. The functional significance of these features is unknown, but they may have been for the siting of specialized musculature associated with chewing. The earliest entelodonts are known from the late Eocene of eastern Asia and thought to be derived from dichobunid stock. *Entelodon* was common in the Oligocene of Europe. While entelodonts were never abundant in North America they did survive there until the early Miocene.

Anthracotheriids

The name anthracothere means 'coal beast'; not animals that ate coal, but so named because many have been found in coal deposits. This aquatic environment of preservation may also have been their life environment, for anthracotheres have many aquatic features. *Merycopotamus*, the last anthracothere in the Plio-Pleistocene of India and Africa, was particularly

Restoration and skeleton of *Entelodon* (length 2 m), from the Oligocene of Europe. The long jaws carried large canine teeth and a formidable grinding dentition behind. The wide zygomatic arch and elongate process suggest very strong jaw musculature. It resembled a wild boar and probably fed on a variety of roots and tubers, supplemented by scavenging. (After Scott).

hippopotamus-like. As the hippopotamus rose, so the anthracotheres declined, an example of ecological niche replacement. Others such as *Anthracotherium* and *Bothriodon* were more pig-like. They had short legs and four-toed feet, but with fusion of paired elements. The skull is elongate with a complete dentition. The incisors and canine are usually normal sized though in some later forms large canines develop. The upper molars are bunodont or bunoselenodont with five main cusps, a feature which makes them easily recognizable. Asia has by far the best record of anthracotheres from mid Eocene times to the Plio-Pleistocene. They are also well known in Europe in the Oligocene and Miocene, and in Africa from the Oligocene onward. Their presence in North America is confined essentially to the Oligocene.

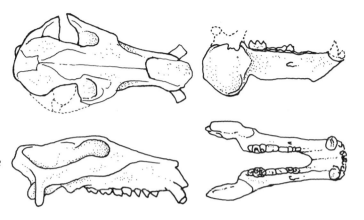

Skull and mandible of *Merycopotamus* (length 28 cm); this is the last known anthracothere and it became extinct during the Pleistocene. The skull is closely reminiscent of a hippopotamus in shape and proportions. (After Colbert).

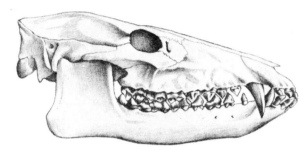

Anthracotherium skull (length 20 cm), a pig-like mammal from the European Oligocene. The upper molar teeth (below) are characterized by having five cusps, three anteriorly and two posteriorly. (After Hünermann).

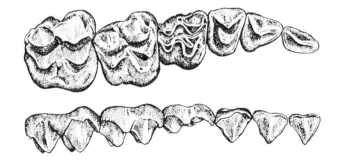

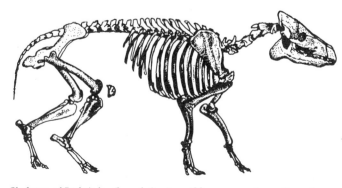

Skeleton of *Bothriodon* (length 2 m), an Oligocene anthracothere from Eurasia. (After Scott)

Pigs and Peccaries

The living Suina or swine comprise the pigs, peccaries and hippopotamuses. The latter have been discussed with the aquatic mammals. The pigs and peccaries are two closely related stocks and it is often difficult to distinguish them in the fossil record. Both can be traced back to Oligocene times, for the most part they are forest rooters and are the ecological successors to the great wave of early Tertiary families that have been described in Chapter 10.

The tayassuids or peccaries are known today in the Americas from Texas to Patagonia where they inhabit terrains varying from near desert to rain forest. They have long slim legs, two-toed feet with very reduced lateral toes and partly fused cannon bone. The snout is pig-like and the face elongate; a small upper vertical canine tusk is often developed. Peccaries reached South America during the Pliocene, but have a record in North America extending back to the Oligocene with for example *Perchoerus*. In Eurasia, peccaries are recorded from the Oligocene and Miocene, and they have recently been described from the Miocene of East Africa and the Pliocene of South West Africa.

The suids or pigs are unknown in the Americas but well represented in the Old World continents from Oligocene times onward. The living babirussa (*Babirussa*), forest hog (*Hylochoerus*) and wild boar (*Sus*) are tropical and temperate forest forms, the water hog (*Potamochoerus*) is semi-aquatic and the wart hog (*Phacochoerus*) a savannah woodland form. Pigs remain four-toed, the lateral toes only slightly smaller than the axial pair. A progressive lengthening of the face is evident, with the development of enlarged upper and often also lower canines; the upper canines grew upward and backward to form great coiled tusks resembling the coiled horns of sheep. *Kubanochoerus* even developed bony outgrowths on the skull above the eyes. The suids basically have low crowned bunodont teeth, often with many small cusplets; the molars of pigs, bears and hominids are all very similar and frequently confused – all three stocks are essentially omnivores.

There are many fossil pigs and because the dentitions are highly variable within each lineage their identification and classification is difficult; it is usual to recognize somewhere between six and ten subfamilies. The pigs probably originated in Asia during Oligocene times and made their appearance in western Europe in the early Miocene. *Conohyus* is a mid Miocene suid known from Pakistan and western Europe. It was a small short-faced pig with complete dentition, enlarged and outwardly turned upper canine, and molars with simple bunodont quadritubercular pattern. The hyotherines ended in the Pliocene with a large specialized pig which had small pointed upper canines – hence its name *Microstonyx*. The babirussa of south-east

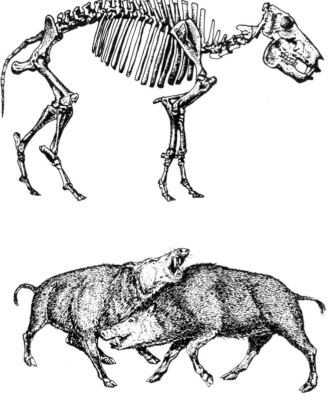

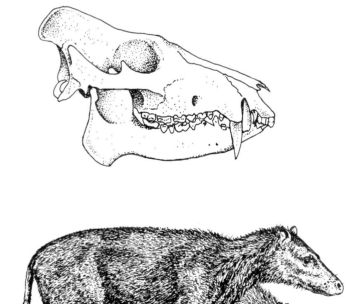

Above, restoration, and top, skull (length 20 cm) of *Perchoerus*, a peccary from the Oligocene of North America showing the well-developed vertical canine teeth characteristic of tayassuids. (After Scott)

Top, skeleton, and above, restoration of *Platygonus* (length 1 m), a tayassuid from the Pleistocene of North America, and recently found still living in Paraguay.

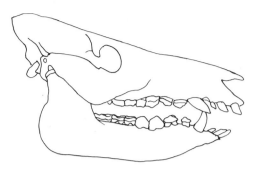

Skull (length 27 cm) of *Conohyus*, from the mid Miocene of Eurasia; an early and unspecialized suid. (After Colbert).

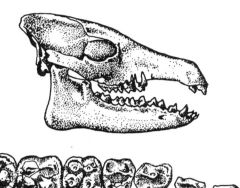

Skull (length 20 cm) and upper dentition of *Palaeochoerus* from the early Miocene of Europe, an early and unspecialized tayassuid. (After Filhol).

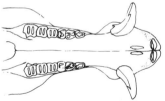 

Palate of *Listriodon*, from the Miocene of Pakistan; this pig has well-developed tusks and each molar tooth has two transverse lophs, thus resembling a tapir molar. (After Colbert).

Skull of *Microstonyx* (length 30 cm) a hypotherine pig from the late Miocene of Europe: the name refers to the very reduced canine teeth. (After Thenius).

Last upper molar tooth of *Metridiochoerus*, seen in side view and occlusal view. Length along occlusal surface 7·5 cm. This Pleistocene pig from Tanzania was a large-sized relative of the warthog. (After Leakey).

Skull of *Kubanochoerus* (length 45 cm) from the Miocene of North Africa; the males of this large pig had a small bony horn above each eye. (After Arambourg).

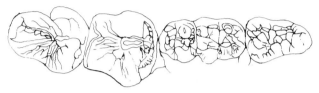

Lower dentition, P_3–M_3 (22 cm) of *Tetraconodon*, a large pig from the Miocene of Pakistan; the two premolar teeth were greatly enlarged. (After Colbert).

Asia is the most primitive living pig. *Listriodon* from the Miocene had large upper canines and developed bunolophodont cheek teeth, perhaps for folivorous feeding; the cheek teeth of *Listriodon* closely resemble those of a tapir, with which they can be confused. *Tetraconodon* in the Pliocene of India has large upper tusks and extremely large and heavy premolars; the dentition has distinct resemblances to that of a hyaena, suggesting it also may have been a bone-crushing carrion feeder. The phacochoerines (warthogs) are the most specialized pigs, and in the Plio-Pleistocene of Africa we can trace their evolution through a series of species over three million years. One of these, *Metridiochoerus*, was very large; it stood over one metre high at the shoulder, had very long curved upper and lower tusks, and the cheek teeth were massive, high crowned, and with very complex enamel patterns. Our domestic pigs can be derived from subspecies of the wild boar, *Sus scrofa*, whose ancestry can be traced back through the Plio-Pleistocene in Eurasia.

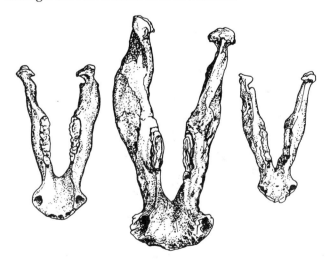

The mandibles of *Phacochoerus* the warthog (right), *Hylochoerus* the giant forest hog (left) – the largest living pig – and *Metridiochoerus* the giant Pleistocene warthog from Tanzania. (After Leakey).

Tylopods

Grouped with the camels and llamas in the Tylopoda are a number of fossil stocks that stand morphologically and probably also physiologically between the primitive Suina and the advanced Pecora. Tylopoda means padded foot, a feature that developed in later camels. Other characteristics of the tylopods are retention in early lineages of swine-like limbs – short, without fusion of bones and little reduced lateral toes; the cheek teeth are primitively bunoselenodont and brachydont. Tylopods have separate histories in the Old World and the New World; a number of families develop in each hemisphere, often paralleling each other. The camelids were the only family to spread successfully worldwide.

Old World tylopods include cainotheres and xiphodonts. Cainotheres occur commonly in the European Oligocene. They were rabbit-sized and very rabbit-like in appearance; with hind limbs longer than the forelimbs they would have bounded or leaped along.

Cainotherium was very advanced in having slender didactyl feet with very reduced lateral toes. The auditory bulla in the ear was large as in desert rodents – suggesting auditory perception was good; also they had large olfactory lobes in the brain indicating a good smell sense. There was a complete dentition, the canine being small and part of the continuous arcade of teeth. The upper molars were five-cusped selenodont teeth. *Cainotherium* may have filled an ecological niche similar to that of rabbits today, but strangely they did not diversify, nor spread outside Europe, nor persist beyond the early Miocene.

The xiphodonts lived in the late Eocene of Europe. *Xiphodon* was about the size of a hare, had long slender legs and was very advanced in being didactyl. The dentition was complete with bunoselenodont five-cusped upper molars, but these differ from those of cainotheres in the arrangement of the cusps; the anterior premolars were very sharp forming a long cutting blade. Xiphodonts bear some striking resemblances to early camelids, but this is probably only due to convergence; for example, in the brain of both lineages the flocculus is well developed (perhaps to enhance balance), but this is also a feature of other primitive artiodactyl lineages.

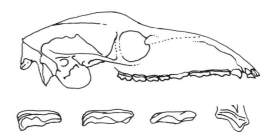

Skull of *Xiphodon* (length 15 cm) and four upper teeth in side view; from left to right P³, P², P¹ and C. *Xiphodon* lived in Europe during late Eocene times and it has some similarities with both cainotheres and early camels. (After Dechaseaux).

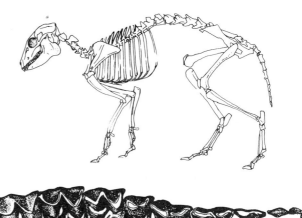

Left, restoration, and above, skeleton (length 30 cm) and dentition of *Cainotherium*, a two-toed rabbit-like herbivore of the European Oligocene that has no known close relatives and no descendants. The molar teeth have five cusps (two anterior and three posterior). (After Hürzeler).

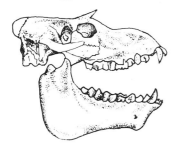

Skulls of oreodonts from the North American Oligocene and Milocene. Top, *Merycoidodon* (length 28 cm), middle, *Leptauchenia* (length 13 cm) and below, *Brachycrus* (length 24 cm). *Brachycrus* was a specialized late oreodont with high nasals suggesting a tapir-like proboscis, as shown in the restoration below. (After Scott, Schultz & Falkenbach).

Among New World families the oreodonts were by far the most successful of the extinct lineages. They were rather pig-like and pig-sized, being heavily built with short four-toed legs and no fusion of the limb bones. The dentition was complete with chisel-shaped anterior teeth. The cheek teeth were strongly brachyselenodont and the upper molars had four cusps. This combination of characters has led to them being dubbed the ruminating swine. Oreodonts roamed the woodlands and grasslands of North America in abundance for nearly 30 million years from the early Oligocene through to late Miocene times. Thousands of *Merycoidodon* have been found in the Badlands of South Dakota, in Oligocene beds referred to as the 'Oreodon Beds'. Unhappily the name *Oreodon* is invalid as a generic and family name and must be replaced by the prior but awkward name *Merycoidodon*; however, the beds may retain the vernacular name of Oreodon. *Merycoidodon* means 'ruminating tooth' and *Oreodon* means 'mountain tooth', an allusion to the mountains in which they are found. Among the specialized oreodonts was *Leptauchenia* from the Oligocene; it had high eyes and nostrils suggesting an amphibious habit like the hippo. Another was the Miocene *Brachycrus* which had a backwardly displaced nasal opening suggesting a tapir-like proboscis may have been present.

Agriochoerus belonged to a small North American family of Oligocene oreodon-like mammals, with long tails and digits which bore claws and not hoofs. The limited mobility of the limb bones and lack of opposable first digit, suggests that they were not true tree climbers but may have clambered about on low branches.

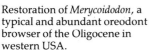

The skeleton (head and body length 1 m) and the front (left) and hind foot of *Agriochoerus*, showing the unusual development of the claws that were used in scrambling along tree branches. These long-tailed herbivorous mammals came from the North American Oligocene. (After Gregory and Scott).

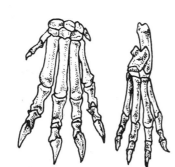

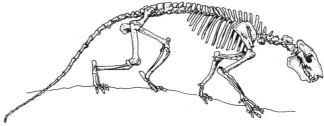

Restoration of *Merycoidodon*, a typical and abundant oreodont browser of the Oligocene in western USA.

Camels

Camels are remarkably well adapted to the extremes of desert environments. There is a record of a caravan of one-humped camels crossing 600 km of Sahara desert in three weeks without water and scant vegetation. In winter, camels can graze for two months without water. In summer, camels can endure desert heat for eight days without water and lose up to 25 per cent of their body weight (a third of their body fluid). Man is in severe stress with a 5 per cent water loss, in delerium with a 10 per cent loss and dead with a 12 per cent loss. As man loses water his blood thickens, this taxes the heart and so slows down circulation; as a result metabolic heat is not dispersed and the internal temperature rises; death follows quickly. A camel avoids death because its blood volume does not change; it has unique crecentric blood cells and it loses water very slowly; its urine is very concentrated. Camels can allow their body temperature to fluctuate over a range of about 7C° (man has about 1C° range). The hump is a fat store, which if oxidised by respiration would yield its weight in water. The stomach has three chambers and water is stored in the rumen. None of all these physiological adaptations is, however, reflected in the skeleton. The surest skeletal sign of desert adaptation is in the feet – the splayed digits with padded foot for sand walking.

From late Eocene through to the end of Miocene times, the camels evolved and diversified exclusively in North America. Then in Plio-Pleistocene times they reached South America where they are represented today by the llama, alpaca, guanaco and vicuña, and they also invaded the Old World where they spread across Eurasia and Africa, surviving today as camel and dromedary. The Eocene camelids such as *Protylopus* were rabbit-sized with four-toed feet and low-crowned teeth. By Oligocene times camels were plentiful in the open woodland habitats of South Dakota; *Poebrotherium* was about goat-sized with complete dentition, and it had lost the lateral toes with the two functional digits already showing signs of diverging.

During the Miocene protolabine camels underwent major transformations coincident with the major expansions of steppe and grasslands in North America. Camels increased in size with lengthening of the neck and limbs; *Protolabis* was about the size of a llama. The cannon bone had formed and the foot dropped from being unguligrade to being digitigrade with the development of foot pads for support in soft terrain. The Miocene camels also adapted to a pacing gait which is a very efficient gait for traversing open habitats, taking long strides with both right and both left legs alternately. There are fossil trackways which show that *Protolabis* was using this gait by mid Miocene times. Associated with these skeletal changes were dental changes; the reduction and loss of the first two upper incisor teeth and their replacement by a horny pad, together with the appearance of procumbent lower incisors, much as in bovids. A diastema formed to separate the front and rear dentitions; the latter changed from low- to high-crowned teeth and so became adapted to feeding on the ever more abundant grasses. There were also a number of sidelines in camel evolution. *Stenomylus* was one of a lineage of small gazelle-like camelids with a peculiar development of some enlarged cheek teeth. Another lineage was that of giraffe-camels with very long legs and necks; these Miocene camelids paralleled the giraffes of the Old World; *Oxydactylus* and *Aepycamelus* were two of the best known of these camels, adapted to browsing on tree foliage out of reach to other ruminants.

Restoration of *Poebrotherium*, an Oligocene camel close to the ancestry of all later camels.

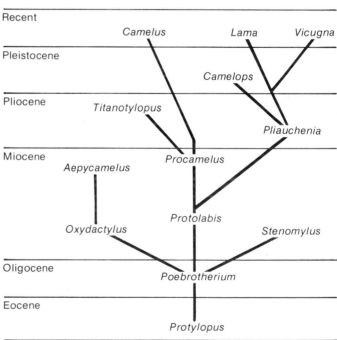

Phylogeny of the Camelidae, showing the principal genera of camels. North America has been the centre of camel evolution throughout the Tertiary and all known pre-Pliocene forms are confined to that continent. Then in Pliocene times they migrated into Asia, Europe, and Africa, and in the early Pleistocene as llamas into South America; they became extinct in their homeland only about 12000 years ago.

The skulls, front feet and molar teeth of three early camel genera. They show the progressive increase in skull (and hence body) size. The foot is small and four-toed in *Protylopus* (skull length 10·5 cm); later it becomes two-toed in *Poebrotherium* (skull length 17·5 cm) and in *Procamelus* (skull length 30 cm) the metacarpals have elongated and fused to form a cannon bone, along with the splaying of the digits which suggests development of a pad on the foot as modern camels to cope with soft terrain. The upper incisor teeth become very reduced; the cheek teeth remain selenodont but become progressively more high crowned until hypsodonty is achieved in *Procamelus*. (After Scott).

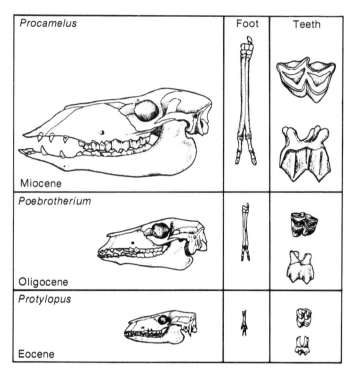

Procamelus	Foot	Teeth
Miocene		
Poebrotherium		
Oligocene		
Protylopus		
Eocene		

Restoration of *Procamelus*, a late Miocene camel with elongate legs and splayed feet.

Restoration of *Aepycamelus* (= *Alticamelus*), a large giraffe-like late Miocene camel that browsed on high tree foliage.

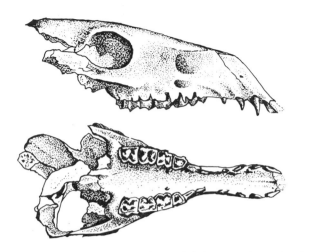

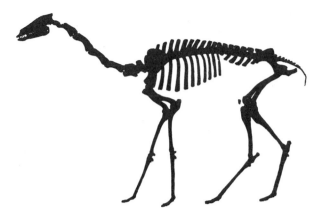

Skull of *Aepycamelus* (length 51 cm); formerly known as *Alticamelus*, this was the largest and last of the giraffe-camels of the Miocene. (After MacDonald).

Skeleton of *Oxydactylus* (length 2.3 m), an early Miocene long legged and long necked camel. (After Peterson).

Restoration of *Oxydactylus*, a tree-browsing camel of the early Miocene scrublands.

In Plio-Pleistocene times there was a further radiation of camelids and invasions of South America and the Old World. Some of these camels were very large; *Titanotylopus* from Nebraska was a giant standing 3·5 m at the shoulder. The South American lineage gave rise to the llamas and the vicuña, all adapted to grazing on the high altitude steppes. In North America *Camelops*, a close relative of the llamas, has been found in abundance in the Rancho la Brea tar pits. It was larger than the living llamas, and contemporary with early man in America; like the horse it had a long evolutionary history on the continent and both became extinct there around 10 000 years ago. However, camels had invaded Asia in the Pliocene and there they still survive. A few reached eastern Europe and Africa, but died out on both those continents. The present camel of Africa is a domestic animal introduced there from Asia around the fifth century BC. The feral camels of Australia are descendants of animals used to cross the Great Australian desert in the 1860s.

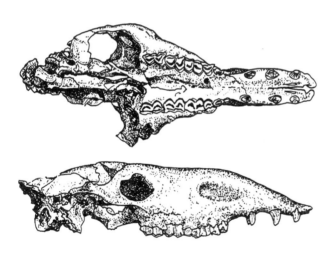

Skull of *Camelops* (length 55 cm), a large llama-like camel which was abundant in western North America during the Pleistocene. (After Webb).

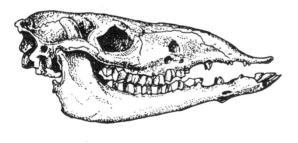

Skull of *Pliauchenia* (length 36 cm), a llama ancestor from the Pliocene. (After Gregory).

Restoration of *Titanotylopus*, a gigantic Pliocene camel of North America which stood 3·5 metres at the shoulder.

Restoration of *Stenomylus*, a gazelle-like camel form the early Miocene.

Protoceratids

Protoceratids can be regarded as a sister group to the
camelids. There are about ten known genera through 55
million years from late Eocene to early Pliocene in North
America, and they appear to have had a preference for
the warmer forested regions in the south. They were
mostly medium-sized ungulates, the size of small deer
such a Virginia deer and Roe deer; they were
conservative in their dental and skeletal features. Most
have low-crowned four-cusped selenodont molars and
only late forms became hypsodont. The metapodials
never fused to form a cannon bone. The feature which
gives the family an indelible stamp is the development
of rostral and frontal horns. The rostral horns may be

Restoration of *Syndyoceras*, an early Miocene protoceratid from North
America with forked nasal horns.

Restoration of *Synthetoceras* from the late Miocene with very large Y-shaped nasal horn.

paired (as in *Kyptoceras*) or fused at the base and forked at the tip (as in *Synthetoceras*). Horns are particularly well developed in males, reduced or absent in females; the bone is dense and in life was probably covered with skin as in giraffes. The horns were visual displays, used offensively and defensively. In primitive protoceratids (as in *Protoceras*) the horns are short and the maximum visual impact is achieved when seen in lateral aspect. In later forms (as in *Synthetoceras*) the maximum visual impact is in full frontal view; this presumably indicates an evolution in sparring behaviour.

Tragulina

Grouped here with the chevrotains are two extinct families and structurally they link the tylopods and the pecorans. They comprise small hornless ruminants, rare in the fossil record and represented today by only two genera, *Tragulus*, the chevrotain of south and east Asia, and *Hyemoschus* the African water chevrotain. Both are small agouti-like forest animals; in the male the upper canines are enlarged into short tusks. *Tragulus* is very similar to one of the earliest tragulids, *Dorcatherium*, well known in the early Miocene of Eurasia and Africa. *Hypertragulus* was a primitive ruminant from the Oligocene of North America and *Archaeomeryx* from the late Eocene of Mongolia is placed in the extinct family Leptomerycidae; the latter had a long tail and an enlarged caniniform lower first premolar.

There are two other families of hornless ruminants which are sometimes placed among the tragulines and sometimes among the pecorans. The gelocids were prominent ruminants in the Oligocene of Eurasia, to be displaced in the Miocene by the moschids, which survive today in *Moschus* the musk deer. Superficially

Head of *Protoceras*, a small forest deer-like protoceratid from the late Oligocene with pairs of small bony horns over the nose, eyes and brain case.

Skull of *Synthetoceras* (length 45 cm); one of the advanced protoceratids about the size of the wapiti or red deer. (After Frick).

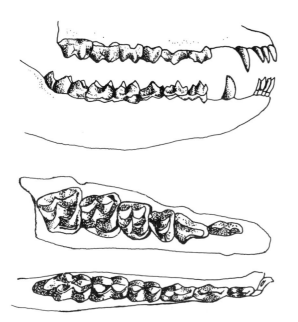

Dentition of *Archaeomeryx* (P_2–M_3 = 35 mm) from the late Eocene of Mongolia. A primitive leptomerycid similar to the living chevrotain; note the caniniform P_1. (After Matthew and Granger).

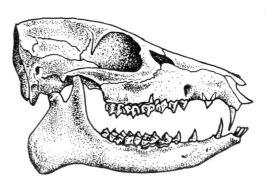

Skull of *Hypertragulus* (length 8 cm), a small hypertragulid from the North American Oligocene. (After Scott).

Upper cheek dentition of *Dorcatherium* (length 3 cm), a tragulid common in the early Miocene of Eurasia and Africa. (After Whitworth).

Skull of *Kyptoceras* (length 45 cm), probably the last survivor of the protoceratids in the latest Miocene of Florida. (After Webb).

Head of *Kyptoceras*, the last protoceratid which died out about 5 million years ago. The orbital horns reached a length of 50 cm.

Skeleton of *Leptomeryx* (length 45 cm), a small chevrotain-like hornless ruminant from the North American Oligocene. (After Scott).

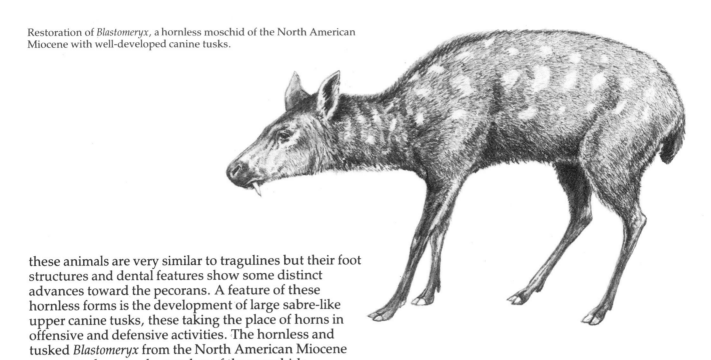

Restoration of *Blastomeryx*, a hornless moschid of the North American Miocene with well-developed canine tusks.

these animals are very similar to tragulines but their foot structures and dental features show some distinct advances toward the pecorans. A feature of these hornless forms is the development of large sabre-like upper canine tusks, these taking the place of horns in offensive and defensive activities. The hornless and tusked *Blastomeryx* from the North American Miocene appears to be an early member of the moschids.

Pecorans

Literally the pecorans are the cattle, but the term is used to include all the higher ruminants – cattle, deer and giraffes. Only a few retain upper canine tusks; in most these are very reduced or absent. The cheek teeth are essentially similar in pattern throughout, which makes identification on teeth alone difficult. The upper molars have four crescentic cusps which vary mainly in height. Deer and giraffes have low-crowned teeth, being woodland and forest browsers; deer in the northern

Opposite: restoration of *Sivatherium*, a gigantic moose-like giraffid that roamed Africa and Asia during the Pleistocene.

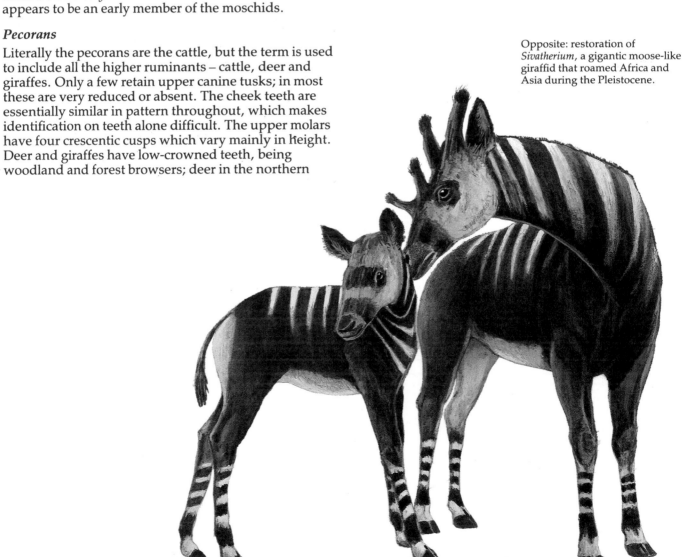

Restoration of *Giraffokeryx*, a palaeotragine giraffid related to the living okapi.

temperate zones and giraffes in the Old World tropics. Antelopes and cattle have high-crowned teeth, being prairie, grassland and savannah grazers; antelopes in the Old World tropics and cattle worldwide in the temperate zones.

In pecorans the ulna bone is reduced and the proximal process fused with the radius; the fibula is likewise reduced to a knob fused to each end of the tibia. The metapodials are fused into a solid cannon bone. The lateral toes are very reduced, but vestiges are usually present except on giraffes. The stomach is fully four-chambered. The most striking feature of the pecora is the development of paired head excrescences; ossicones in giraffes, antlers in deer and horns in cattle.

GIRAFFIDS   The first percorans appear in the early Miocene of Europe and Africa and are difficult to assign to family status, hence the origin of the three main lineages (giraffes, deer and cattle) remains obscure.

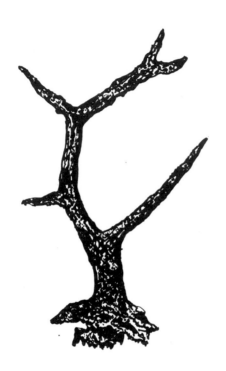

Ossicone of *Climacoceras* from the mid Miocene of East Africa. This branched bony outgrowth lacks a burr and therefore fits the definition of an ossicone; dentitions associated with it at Fort Ternan in Kenya have giraffoid features, e.g. a bicuspid lower canine. (After Hamilton).

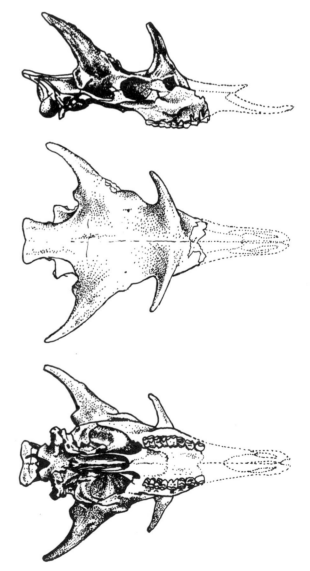

Skull of *Samotherium* (length 54 cm) from the late Miocene of Eurasia and Africa. This palaeotragine giraffid has a pair of ossicones on the frontal bones above the orbits with often a second rudimentary pair anterior to them. (After Lydekker).

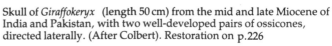

Skull of *Giraffokeryx* (length 50 cm) from the mid and late Miocene of India and Pakistan, with two well-developed pairs of ossicones, directed laterally. (After Colbert). Restoration on p.226

Climacoceras from the Miocene of East Africa has well-developed branched ossicones – the absence of a burr indicates they were not shed and so suggests giraffoid affinities. Giraffids were abundant and varied during the Miocene and Pliocene in the Old World. As well as primitive forms like *Palaeotragus* and *Samotherium* with a pair of unbranched ossicones, there was *Giraffokeryx* from India with two pairs of ossicones and *Sivatherium*, a gigantic Pleistocene giraffe also from India with a large palmated ossicone, with heavily rugosed enamel on the cheek teeth and with a short proboscis; in life it must have resembled a moose. The giraffids are forest browsers, some in equatorial forests as the okapi and others in woodland savannah as the giraffe, which uses its great reach to feed on foliage high up on trees. The sivatheres appear to have persisted into early Recent times; there are petroglyphs of them in the Sahara and a bronze figurine from ancient Sumeria.

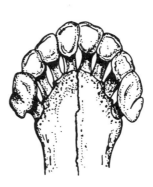

Mandibular symphysis of *Giraffa* (giraffe) showing three pairs of incisor teeth with alongside a pair of bilobed canine teeth. Maximum width across canines 10 cm.

Top: skull of *Sivatherium* from the Pleistocene of India. This gigantic giraffid skull (length 70 cm) had in male specimens a conical pair of ossicones above the orbits and large palmate pair behind on the parietal bones. Restoration on p.227
Above: Upper dentition of *Sivatherium* (length 25 cm) showing the molarized premolars and the hypsodont molars with thick rugose enamel. (After Churcher).

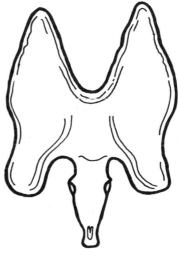

Skull of *Prolibytherium*, an early sivatherine giraffid with a pair of large palmate ossicones (maximum span 35 cm). Early Miocene of North Africa.

Mandibular cheek dentition of *Giraffokeryx* (length 14 cm). (After Ciric & Thenius).

Upper cheek dentition of *Palaeotragus* (length 14 cm), a primitive giraffoid from the mid Miocene of East Africa. (After Hamilton).

Petroglyph of a possible sivathere from a rock shelter in the Tibesti mountains, central Sahara. A contemporary representation of a fauna living some 8000 or more years ago.

CERVIDS   The small deer *Euprox* was common in mid Miocene times in Europe; with its short pointed antlers on long bony pedicles it must have closely resembled a muntjac, the most primitive living deer. In the course of evolution cervids have lost their upper canine tusks, shortened the pedicle and greatly enlarged the antlers. They have become the dominant herbivores of the northern temperate forests and woodlands. They have diversified into many niches, including semi-aquatic forms like the moose or elk (*Alces*) and the Asiatic water deer, and into the tundra with reindeer (*Rangifer*), in which both sexes have antlers. Cervids never successfully invaded Africa though they did in Plio-Pleistocene times penetrate North Africa. The genus *Cervus* includes the red deer of Europe and the wapiti of North America.

Restoration of *Eucladoceros*, a large male deer with twelve tines or points on each antler. The name *Eucladoceros* means 'well branched antler'. Span 1·7 metres.

Euprox from the early mid Miocene of Europe. This small antlered mammal is one of the earliest known cervids; the bony pedicle with burr and short pointed antler (total height 8 cm) closely resemble the living muntjac deer. (After Thenius).

Head of *Procranioceros*, a three antlered deer from the mid Miocene of North America. Each antler is short branched and carried on a bony pedicle.

One of the most spectacular deer was the European Pleistocene bush-antlered *Eucladoceros* with a positive forest of tines. The most celebrated fossil deer is undoubtedly *Megaloceros*, sometimes misnamed the Giant Irish Elk. Giant it is, but it is not an elk, being essentially an outsized fallow deer. Further it is not exclusively Irish; its range during the Pleistocene extended eastward from Ireland through Siberia to China; many skeletons have, however, been recorded from Irish bogs. The large palmate antlers in a mature male could span 3·7 metres and weigh around 45 kg, which is about one seventh of the deer's weight. At sexual maturity, a stag will cease to grow larger, but each year it will produce larger antlers, so that the older stags carry the heaviest crowns, and it is this allometric growth that is seen by some as the cause of their extinction. Seeing a stag head on, it must have presented a formidable impression and served the

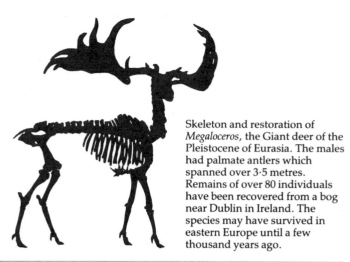

Skeleton and restoration of *Megaloceros*, the Giant deer of the Pleistocene of Eurasia. The males had palmate antlers which spanned over 3·5 metres. Remains of over 80 individuals have been recovered from a bog near Dublin in Ireland. The species may have survived in eastern Europe until a few thousand years ago.

owner well in enabling him to achieve and maintain dominance over a herd. Giant deer flourished in open woodlands during a warm interval in late glacial times, but with the onset of a cold spell about 12 000 years ago their numbers were drastically depleted, though a few appear to have held out in Austria and around the Black Sea until about 500 BC, and are by some identified with the 'Schelch' of the Nibelungen Lied.

A recent discovery at Gargano in Italy, in the same Miocene deposit which yielded the giant hedgehog (see p. 49) is of a bizarre five-'horned' ruminant with sabre-like canine tusks; *Hoplitomeryx* is so unlike any known ruminant it is impossible to place it with either deer or cattle.

BOVOIDS   These comprise two families, both going back to Miocene times, the antilocaprids in North America and the bovids in the Old World. Today *Antilocapra* (the prongbuck) is the only living member of the family and it ranges from Canada to Mexico. *Merycodus* was a Miocene genus and in some ways comparable to antelopes. *Merycodus* horns were present in both sexes and were forked with a burr at the end of the pedicle and the outer horn was shed annually as in the prongbuck. In the Plio-Pleistocene there was a vast array of antilocaprids, many with bizarre adornments.

The earliest bovids appear in the early Miocene of the Old World; from the Sahara come horn cores of *Eotragus* and some other small gazelle-like bovids, and similar ruminants are recorded from beds of about the same age in Mongolia and France. From these beginnings, around 20 million years ago, with small horned ruminants living in woodland savannahs, there radiated during mid Miocene times a modest 15 new genera, most of them in Asia. But by late Miocene times 10 million years ago we have records of 70 new bovid genera, a veritable explosion of new types. These are mostly cursorial, able

to move rapidly over the open plains, and have high-crowned teeth able to cope with tough grasses. By Pleistocene times there were over 100 genera, but it has now dropped to around half that figure.

The most distinguishing feature of bovids is the size and shape of their horns; with only the cores fossilizing, much valuable information is lost. Horns developed in most ruminants as they reached a threshold size of around 15–20 kg. The horns are displayed for offense and defense, and their development relates closely to

Skull and head of *Hoplitomeryx*, (skull length *c.* 18 cm) a strange late Miocene ruminant from Italy with sabre-like canine teeth and six bony skull outgrowths; it is not known whether this animal is a cervid, bovid or a representative of an otherwise unknown family of ruminants. (After Leinders).

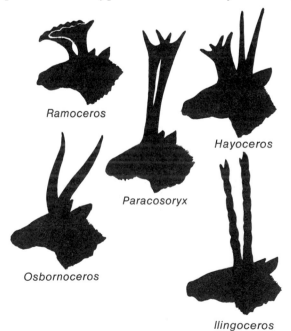

Ramoceros

Hayoceros

Paracosoryx

Osbornoceros

Ilingoceros

Heads of a series of fossil antilocaprids from North America to illustrate some of the diversity of horn patterns. *Ramoceros* (mid Miocene); *Paracosoryx* (mid Miocene); *Osbornoceros* and *Ilingoceros* (late Miocene); *Hayoceros* (Pleistocene). (after Frick).

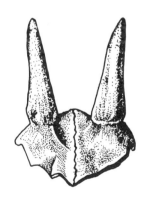

Horn core of *Eotragus* (core length 8 cm), a small primitive bovid form the mid Miocene of Europe. (After Filhol).

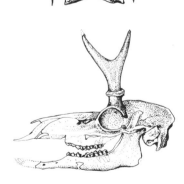

Skull of *Merycodus* (length 18 cm) an antilocaprid from the late Miocene of North America. This prongbuck has a branched pair of 'horns' each with a basal 'burr'. (After Frick).

Head of *Hexameryx*, (skull length 21 cm) a strange six 'horned' antilocaprid from the late Miocene of Florida. (After Webb).

SOME LIVING BOVOIDS

FAMILY	SUBFAMILY	TRIBE	EXAMPLE
Antilocapridae			Pronghorn
Bovidae	Bovinae	Tragelephini	Bushbuck
		Boselaphini	Indian nilgai
			Fourhorned antelope
		Bovini	Cattle
			Bison
			African buffalo
			Indian water buffalo
	Cephalophinae	Cephalophini	Duikers
	Hippotraginae	Reduncini	Reedbuck
		Hippotragini	Oryx
			Roan antelope
			Sable antelope
	Alcelaphinae	Alcelaphini	Gnu
			Impala
	Antilopinae	Neotragini	Dik dik
		Antilopini	Gazelles
	Caprinae	Saigini	Saiga
		Ovibovini	Muskox
			Takin
		Caprini	Goats
			Sheep

the evolution of social structure, territoriality, feeding habits and predator pressures. While the family appears to have had its major centre of evolution in tropical regions of the Old World, some members have adapted to deserts (e.g. oryx, addax and Dorcas gazelle) and some to semi-aquatic environments (e.g. reedbuck and water buffalo). Only in mid Pleistocene times do bovids appear to have penetrated cold northern latitudes and migrated across Beringia into the New World, where today they live on as bison, muskox, bighorn sheep and mountain goat. They never reached South America, but the bison did reach El Salvador.

Restoration of *Ilingoceros*, an antilocaprid from the late Miocene of North America.

Restoration of *Pelorovis*, a gigantic relative of the African buffalo. *Pelorovis* was one of several herbivores (along with pigs, giraffes and deinotheres) that achieved gigantic size in the East African Pleistocene.

Restoration of *Tsaidamotherium* from the late Miocene of Mongolia; this unusual bovid has asymmetrical horn cores; the left is vestigial and the right is large and medianly placed.

The family can be divided into six subfamilies and twelve tribes; none is extinct and in most cases the fossils show many similarities to the living representatives. *Tsaidamotherium* from the latest Miocene of Mongolia was probably a relative of the muskox. It had very unequal sized horns; the left core was atrophied and the right one formed a large centrally positioned horn base. Of other extinct genera *Pelorovis* is outstanding in size; it was a gigantic African Pleistocene bovid related to the African buffalo *Syncerus*. The horn cores spanned 2 metres and each individual horn may have reached 2 metres in length. It only became extinct about 12 000 years ago. Our domestic cattle are mostly derived from aurochs (*Bos primigenius*), which only became extinct in the seventeenth century, having been first domesticated as early as the fourth millennium BC.

Skull of *Pelorovis* below that of a living *Syncerus*, African buffalo. In *Syncerus*, the right horn is in place while the left side displays only the horn core for direct comparison with *Pelorovis* horn cores. (After Leakey).

Cave painting of the aurochs, *Bos primigenius*. Upper Palaeolithic period; Lascaux Cave, France.

Cave painting of the steppe wisent, *Bison priscus* from the Upper Palaeolithic period; Font de Gaume, France.

Above: restoration of the giant bison, *Bison latifrons* from the late
Pleistocene of southeast USA. This great bison had horn cores that
spanned over 200 cm, compared with about 65 cm for the living North
American bison.

Below: restoration of *Bos primigenius*, the wild cattle of the late
Pleistocene and in part ancestral to our domestic cattle.

CHAPTER 13
Of men and monkeys

'Many that are first shall be last and the last shall be first'
MATTHEW 19:30

Linnaeus opened his great *Systema Naturae* on the classification of organisms with the mammals and amongst these he placed first the order Primates; in the premier order are placed lemurs, monkeys, apes and man, a grouping of about 60 living genera and about twice that number of fossil genera. Man excepted, primates are today and have been throughout their history, essentially tropical forest dwellers. They are scansorial, small- to medium-sized vegetarians or omnivores. The smallest is the pygmy marmoset weighing around 70 g; only gorillas exceed man in weight. Primates have five digits on each foot; these usually bear flat nails, though a few have claws. Climbing adaptations seen in almost all primates include great freedom of movement in the limb joints, ability to rotate the radius on the ulna and an opposable first digit on both front and hind foot. Many have long tails and in some of these they are prehensile and used as a fifth limb in climbing. The dentition usually has two pairs of transverse spatulate incisor teeth. The canines vary and are often larger in the males. The cheek teeth are normally low crowned with rounded cusps, the back teeth often with accessory cusps. The brain is proportionately large for the size of the species and especially large in apes and man. Smell is often poor, but hearing is good and vision generally stereoscopic. Most primates are highly vocal.

When we take this roster of primate features, we can see that man differs from other primates in a number of details, but not in fundamentals. He has moved out of the tropical forests to inhabit almost every terrain on earth. He has become totally bipedal and lacks an opposable first toe. The upright posture is reflected in

PRIMATES

Order PRIMATES		
Suborder	*Plesiadapiformes*	
Family	Paromomyidae	† *Purgatorius*
	Plesiadapidae	† *Plesiadapis*
Suborder	*Strepsirhini*	
Family	Adapidae	† *Smilodectes*
		† *Azibius*
		† *Lushius*
	Lemuridae	† *Megaladapis*
Suborder	*Haplorhini*	
Family	Omomyidae	† *Tetonius*
		† *Necrolemur*
	Tarsiidae	*Tarsius*
Suborder	*Platyrrhini*	
Family	Cebidae	† *Branisella*
Suborder	*Catarrhini*	
Superfamily & Family	Parapithecidae	† *Parapithecus*
Superfamily & Family	Cercopithecidae	† *Dinopithecus*
Superfamily	Hominoidea	
Family	Pliopithecidae	† *Propliopithecus*
Family	Pongidae	† *Dryopithecus*
		† *Gigantopithecus*
		† *Sivapithecus*
		† *Ramapithecus*
Family	Hominidae	† *Australopithecus*
		Homo

† = Extinct

many skeletal changes in the vertebral column, hip bone, limbs and feet. In modern man the brain size averages around 1360 cm^3. This gives him the highest known brain/body weight ratio.

	Brain weight kg	Body weight kg	Brain weight × 100 / Body weight
Man	1·36	55·5	2·45
Gorilla	0·57	172·4	0·33
Elephant	5·7	6700	0·09
Sperm Whale	9·2	36000	0·03

Other features which distinguish man include social and behavioural practices, none of which fossilizes, though information can sometimes be gleaned from the fossil record. The implements used by Stone Age man reveal much about his ways of life. Indeed the ability to fashion and use implements has been thought to be unique to man and hence the conclusion that man could be defined as the 'tool-maker'; however, many animals utilize natural materials to build dwellings and obtain food. It was Aristotle who defined man as the rational

Ring tailed lemur (*Lemur catta*). Lemurs are the most primitive living primates and confined to the forests of Madagascar where they feed on vegetables and fruits. (After Wilson).

The two major groups of monkeys can be distinguished by their nostrils. In the Old World monkeys (top) or catarrhines (e.g. *Macaca* the macaques) the nostrils are separated only by a thin vertical partition. In the New World monkeys (above) or platyrrhnes (e.g. *Cebus* the capuchins) the two nostrils are quite separate.

The four living ape genera are shown here. *Hylobates* the gibbon (top right); *Pongo* the orang-utan (top left juvenile and top middle adult face); *Pan* the chimpanzee (lower left); *Gorilla* the gorilla (lower right). All are tailless; the chimpanzee and gorilla are confined to tropical Africa, the gibbons and orang-utan to south-east Asia.

The black capped capuchin (*Cebus apella*) is an example of the cebids, monkeys which inhabit tropical forests of Central and South America. (After Wilson).

The olive baboon (*Papio anubis*) is an example of the cercopithecids, monkeys which inhabit warm temperate and tropical regions of Africa and Asia. (After Wilson).

animal – he had animality and rationality. This may be true but is not applicable to fossils. There is no one clear character or state which would be identifiable in a fossil and which would enable us to say it was human. Bipedal posture and large brain were acquired gradually over a long period. The most we can do is trace the fossils through time and recognize where stocks diverge from each other and hopefully identify that from which man emerged. Being forest dwellers, the fossil record of primates is not good, but by dint of enormous expenditures of time and money, particularly over the past three decades, we now can piece together many steps in the evolutionary sequence.

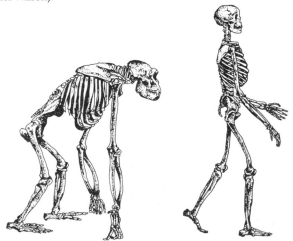

Skeleton of *Gorilla* (left) and *Homo sapiens* (right). The elongate pelvis and long front arms of the gorilla help differentiate the skeleton from that of man.

Early forms

The first primate

The earliest supposed primate is *Purgatorius* from the
Rocky mountains. The name is based on one isolated
lower molar with a trigonid of cusps and a broad talonid.
From slightly higher beds in the Palaeocene come good
dentitions, showing the upper molars to have a triangle
of cusps and a broad inner shelf. *Purgatorius* was about
the size of a small rat and indeed may have had a rat-like
life. At one time early primates were thought to be
closely related to hedgehog ancestors, but differences in
their basal skull bones make this most unlikely; it could
be that primates and rodents share a common ancestor
back in the late Cretaceous, but evidence is still very
slender.

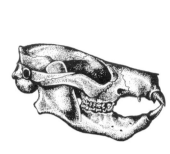

Restoration and skull (length
10 cm) of *Plesiadapis*, a 55-million-
year-old squirrel-like primate
which inhabited forests of North
America and Europe. Note the
rodent-like incisors, followed by a
long diastema. (After Simons).

Upper molar tooth of *Purgatorius*,
the earliest known primate genus
from the late Cretaceous and
early Palaeocene of western USA.

The plesiadapids

In the Palaeocene of western North America and
western Europe we know a series of primates with close
affinities to paromomyids, the family which includes
Purgatorius. Of these the plesiadapids are the best
known. *Plesiadapis* from France was an arboreal primate
with claws, a good sense of smell and laterally placed
eyes – indeed very squirrel-like in appearance and
occupying an ecological niche similar to that of squirrels
today.

Adapids and omomyids

By Eocene times two new major families had appeared,
the adapids and the omomyids; both were abundant and
varied, and show considerable advances on the
plesiadapids. *Smilodectes* from the North American
Eocene was a lemur-like adapid, with long hind limbs, a
large brain, a blunt snout and stereoscopic vision. It was
from close relatives of the adapids that the lemurs and
lorises are descended. Also in the Eocene faunas were
the omomyids, another highly varied assemblage of
small arboreal primates whose nearest living relative is
the tarsier, the tiny nocturnal insectivorous primate in
the forests of south-east Asia with very large eyes and
the ability to turn its head through almost 360°. Two
such *Tarsius*-like Eocene omomyids were *Tetonius* from
North America and *Necrolemur* from western Europe.
Primates are unknown from both Africa and Asia during
the Palaeocene, but from the Eocene there are
tantalizing fragments which are difficult to assign to
family status; *Azibius* from Algeria could be an adapid,
as also might be *Lushius* from China. However, it is the
status of two Burmese taxa, *Amphipithecus* and
Pondaungia which arouses considerable interest; they
have been thought to be primitive catarrhini, which
places them at the base of the stock that gave rise to
monkeys, apes and hominids. However, they are very
poorly known genera and until more material is
available, relationships remain highly conjectural.

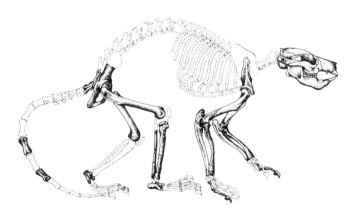

Reconstructed skeleton of *Plesiadapis*. Note the long tail, long limbs,
completely separate ulna/radius and tibia/fibula enabling the feet to be
turned and with the elongate digitis enabling them to grasp branches.
(After Simons).

Restoration (opposite) and skull
(length 6 cm) of *Smilodectes*, a 50-
million-year-old primate from
western USA. The long tail,
elongate feet and opposable first
digits are all adaptations to
climbing. The short face,
complete dentition without a
diastema and binocular vision
give this primate a very lemur-
like appearance. (After Simons).

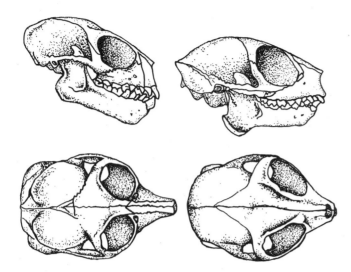

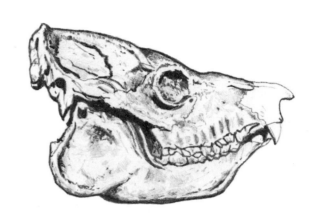

Lemurs and lorises

The nearest living relatives of these Eocene stocks are the lemurs of Madagascar and the lorises of Africa and southern Asia. The strangest of all lemurs is the recently extinct *Megaladapis* from Madagascar. This gigantic lemur had a large skull 30 cm long and a body weight of around 50 kg. It was an adept tree climber and probably a foliage eater; in these two respects it would have resembled the living koala of Australia.

Skulls of two small tarsioid primates from the Eocene; *Tetonius* (right) from western USA and *Necrolemur* (left) from western Europe. Skulls about 3·5 cm long. (After Szalay & Cleary).

Upper and lower dentition of *Tetonius* from Eocene of western USA (length M³ − I¹ = 16 mm). Note the small canine tooth, loss of anterior premolar and broad crushing molars. (After Szalay).

Megaladopsis, top: skull (length 30 cm, after Tattersall), middle: Silhouette of skeleton (length head and body 1·5 m) and, bottom: restoration. This giant leaf-eating lemuroid lived in the forests of Madagascar until recently. It was a browsing tree dweller which resembled an outsized koala, being a powerful climber rather than a brachiating acrobat.

Advanced primates

Our knowledge of primate evolution over the crucial ten million years of Oligocene time is extremely slender. Aside from a few surviving omomyids in western North America and Europe, we have evidence from only two areas, South America and North Africa; in these regions we see the appearance of the two major post-omomyid primate stocks, platyrrhines and catarrhines. These two stocks are differentiated geographically and anatomically. Platyrrhines occur in the New World (Central and South America), have flat noses with paired but well-separated outwardly directed nasal openings, they retain three premolar teeth in both jaws and the anterior upper molars have three or four major cusps; they comprise the cebid monkeys and marmosets. The catarrhines occur in the Old World (Eurasia and Africa), have the paired downwardly directed nasal openings closely appressed, usually have only two premolar teeth in each jaw and the anterior upper molars always have four major cusps; they comprise cercopithecid monkeys, apes and man.

The New World platyrrhines – marmosets and cebid monkeys

From the early Oligocene of Bolivia we know *Branisella*, the first primate on the continent, and the earliest and most primitive known platyrrhine. In the late Oligocene of Patagonia are several new platyrrhine genera, including marmosets. *Branisella* retains many omomyid features – but overall its dental characters have affinities with the cebid monkeys, which suggests it is close to platyrrhine ancestry.

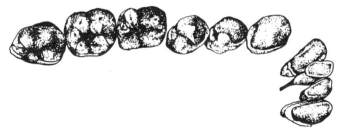

Lower dentition (P$_2$–M$_3$ = 22 mm) of *Parapithecus*; this small primate is one of the earliest known catarrhine monkeys.

The Old World catarrhines – cercopithecid monkeys and the apes

From the early Oligocene of the Fayum in Egypt we have a fauna of catarrhines which fall mainly into two families – the parapithecids and the pliopithecids. *Parapithecus* primitively retains three premolars, but its upper molars have four well-developed cusps in the catarrhine pattern; its many derived features, however, suggest it may be a sideline rather than an ancestor to later stocks. Another Fayum primate is *Propliopithecus*, which is well represented with dentitions and skull material. It was a

Upper molar dentition of *Branisella* (M$^{1-3}$ = 8 mm), the earliest known South American primate from the early Oligocene of Bolivia.

Head and mandibular dentition (P$_4$–M$_3$ = 19 mm) of *Propliothecus*, a spider monkey-like pliopithecid from the Oligocene of Egypt. The large eyes gave full stereoscopic vision and the bulbous-cusped bunodont molars suggest a frugivorous diet. *Propliopithecus* is probably close to the ancestry of apes and Old World monkeys.

small-sized monkey-like animal with a dentition that resembles a primitive ape. The canines are large, the lower premolars sectorial and the molars with beaded cingula. While not a true pongid or ape, it could be a member of a stock that gave rise to both pongids and hominids.

THE CERCOPITHECID MONKEYS   In Miocene sequences in the Old World we can trace the radiation of two superfamilies of catarrhines, the cercopithecoids and the hominoids. The cercopithecoid monkeys are found through much of Africa and southern Asia and have reached as far north as England (in the Pleistocene of Grays Thurrock in Essex) and Japan. They were fairly abundant in the Plio-Pleistocene of Africa where several large baboon species are known (e.g. *Theropithecus*). Human interest, however, tends to focus on the divergence of hominoid families, of which the two main lineages are the pongids (apes) and the hominids (humans).

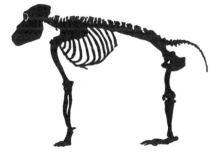

Reconstructed skeleton of *Theropithecus* from the Pleistocene of the Olduvai Gorge in Tanzania. This large baboon, a close relative of the living Ethiopian gelada, stood 75 cm at the shoulder. The face is short and deep for a baboon, the strong sagittal crest on the skull roof indiates powerful masticatory muscles. It was probably a ground dweller feeding off tough vegetation on the savannah.

THE HOMINOIDS   The pongids today comprise the
gibbons, orang-utan, chimpanzee and gorilla. The
gibbons of south-east Asia are universally recognized as
divergent from the remaining apes. The fossil pongids
are numerous in specimens, and oversplitting has
added many names to the literature; they are also widely
distributed across Africa and Eurasia. There is little
agreement on which is closest to man, it depending
greatly on which pieces of evidence are espoused and
which rejected. We can reasonably recognize four fossil
genera which have varying degrees of pongid and/or
hominid features; they are *Dryopithecus*, *Gigantopithecus*,
Sivapithecus and *Ramapithecus*. *Pithecus* means 'ape' in
Greek, and the first two suffixes mean tree and gigantic.
The latter two from Pakistan are named after the Hindu
gods Siva and Rama.

  Dryopithecus is relatively abundant in Miocene faunas
of Africa and in late Miocene faunas of the
Mediterranean area, with at least four subgenera and
about eight species. *Dryopithecus* (*Proconsul*) *africanus*
from the early Miocene of Kenya is one of the best
known species. It was about the size of a rhesus
monkey, the canines are prominent but smaller than in
living apes; the cheek teeth are proportionately small
and have thin enamel, thus wearing down rapidly.
There are no brow ridges, the jaw projects moderately
and the brain is gibbon-like. The postcranial skeleton
has many similarities with monkeys and gibbons, and
Dryopithecus was undoubtedly a quadruped. It is not
known to share any specialized characters with living
apes. The second genus *Gigantopithecus* is known from
Pakistan, India and southern China in beds ranging
from late Miocene into the Pleistocene. It was a gorilla-
sized ground ape with robust canines and molarized
premolars. The molar teeth have thick enamel, high
crowns with low cusp relief and often accessory
cuspules. *Sivapithecus* is known from mid and late
Miocene deposits of southern Europe, Asia and East
Africa, with around four species; the canines are of
moderate size and the molars have thick enamel.
Sivapithecus is found in assemblages of open country or
mixed habitat rather than with forest faunas; this may

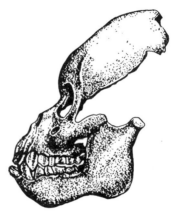

Head and skull of *Proconsul*. This
dryopithecine pongid was about
the size of a rhesus monkey and
common in Miocene times in East
Africa and southern Europe
where various sized species
inhabited tropical and warm
temperate open woodlands.

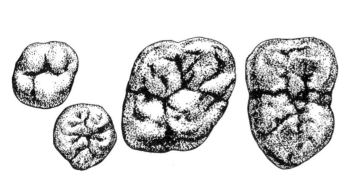

Head of *Gigantopithecus*, a Pleistocene ape from China. This gorilla-
sized pongid lived in wooded savannahs. It was first known from
isolated teeth obtained from Chinese drug stores, as were the above,
teeth shown here (right) beside those of modern man for size
comparison.

Restoration of *Ramapithecus* from the mid Miocene of Pakistan. A quadrupedal pongid with a number of hominid dental characteristics, it may be close to the ancestry of hominids. The robust jaws suggest it could crush and grind tough food gathered from mixed woodland and savannah habitats.

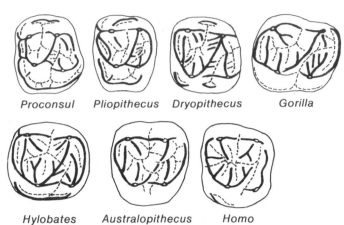

Proconsul   Pliopithecus   Dryopithecus   Gorilla

Hylobates   Australopithecus   Homo

Upper molar teeth of a series of hominoid primates. All drawn to the same transverse width, they show the relationships of the major cusps. (After Butler & Mills).

suggest it was more of a ground dweller, feeding on coarse or tough foodstuffs, the thick enamel reducing the rate of wear. The fourth genus *Ramapithecus* has a similar space and time distribution to *Sivapithecus*; they often occur together in the same sites and are difficult to distinguish. *Ramapithecus* has large thick enamelled cheek teeth with high crowns and the canines are low as in early hominids, though a dental arcade without a diastema has not formed. *Ramapithecus* retains many pongid characters and has some hominid characters which lead many writers to consider it the closest of the four genera to the hominids. This, however, may be more a reflection of adaptation to a food niche similar to that of early man, rather than a true affinity.

Following the appearance of *Sivapithecus* and *Ramapithecus* some 14 million years ago in Kenya and Pakistan, there is a surprising hiatus with virtually no knowledge from anywhere in the world of new higher primates until around 4 million years ago. When we again pick up the record after the 10-million-year gap it is a very different assemblage of primates. During this time gap much evolutionary change occurred among the mammals; it represents the span between the early mastodonts and the first elephants, between the first grazing horses and the first one-toed horses. When we again see hominids in the Pliocene of Africa they are very different from the Miocene *Ramapithecus* and its allies.

There are numerous sites in East Africa which yield Plio-Pleistocene hominids of between one and four million years old. Laetoli near Olduvai Gorge in

Tanzania and Hadar in central Ethiopia are among the best known. Only slightly younger successions are well documented in the Omo valley in southern Ethiopia, around the eastern shores of Lake Turkana in northern Kenya, in Olduvai Gorge in Tanzania and in cave sites on the Transvaal in South Africa. From these and other sites in Africa, southern Europe and Asia, we can distinguish two genera of hominids, *Australopithecus* and *Homo*; each has a bewildering range of morphotypes which, with the small population samples available, makes species identification horrendously difficult and the source of endless lively debates.

Both *Australopithecus* and *Homo* differ from all other hominoids in having upright postures and a dentition arranged in a parabolic arcade without a diastema between upper incisors and canine. Gorilla brains average about 500 cm^3 with the largest approaching 700 cm^3. *Australopithecus* brain is of similar size but it had different proportions and the cerebral cortex pattern was much more like that of man. Human babies achieve speech when their brains have reached about 600 cm^3, which is just above the size of that of an *Australopithecus* adult, so it is perhaps unlikely that the southern apes had speech. The tools found at australopithecine sites cannot with certainty be attributed to them as often they were accompanied by *Homo habilis*.

Australopithecus was first found in caves in the Transvaal in the 1920s. Over the decades a rich assemblage of specimens and their associated faunas has been recovered from sites that range in age from one to nearly four million years. Other australopithecines have been found in East Africa, but none from outside Africa. There are two distinct species of *Australopithecus* recognizable – *A. africanus* and *A. robustus*; they differ in height, weight and brain capacity, and in the crown areas of their cheek dentitions.

The earliest dated australopithecines come from sites at Laetoli in Tanzania (3.7 million years), Hadar in Ethiopia (3.0 MY) and Transvaal (2.5–3.1 MY). These australopithecines may include several species and subspecies or they may all belong to the same species, albeit a rather variable and long lived species. Professor

Tobias regards them all as *Australopithecus africanus*, with each locality yielding a different subspecies, thus:

  A. a. africanus for the Taung skull from Cape Province
  A. a. transvaalensis for gracile specimens from Sterkfontein and Makapansgat
  A. a. tanzaniensis for the Laetoli specimens
  A. a. ethiopicus for the Hadar specimens.

The summed crown areas of the cheek teeth (P_3–M_3) show the Laetoli specimens to be close to those of the Transvaal, but the Hadar specimens have very much smaller teeth.

Australopithecus robustus (3 morphs)	882–960–1312
'Laetoli' morph	869
Australopithecus africanus transvaalensis	861
Homo habilis	787
'Hadar' morph	733
Homo erectus (4 morphs)	544–695
Homo sapiens	485

Table of crown areas of cheek teeth (P_3–M_3) in mm^2. (After Tobias.)

Restoration of *Australopithecus robustus*. This hominid lived during the early Pleistocene in eastern and southern Africa in open savannah woodland, feeding mainly on fruit and nuts.

The second hominid genus *Homo* has three distinct species. The earliest is *H. habilis*, a small species which has many close similarities with *A. africanus*. However, there are some distinct differences in the skull bones and the cheek teeth are smaller than those of all the australopithecines save for the Hadar fossils. *H. habilis* remains have been found by Lake Turkana in northern Kenya, in the Olduvai Gorge of Tanzania and probably also in the Transvaal; it survived for around two million years, being replaced by *H. erectus*. The cheek dentition of *H. habilis* is fairly similar to that of *A. africanus*, though smaller and generally with thinner enamel. The brain capacity of *H. habilis* was larger, though not the stature; it walked upright and had a powerful grasping hand well suited to swinging through trees. We can thus envisage a life style not dissimilar to that of baboons today – ground foraging by day and spending the night safely in trees. The Olduvan tools associated with at least some of the fossils, if truly the artifacts of *H. habilis*, might be interpreted to indicate use of animal products, though not necessarily primary hunting of animals.

The second *Homo* species we recognize is *Homo erectus*. This species was the first member of the genus to leave Africa in the early Pleistocene; his remains are known across Africa from the Cape to Morocco, from Chad to Kenya, in Europe (Hungary), China (Peking Man), in south-east Asia (Java Man). *H. erectus* was larger than *H. habilis* and had a much larger brain, a brain that reached the size of modern man; his teeth were very similar to those of modern man. He had an improved stone culture and had harnassed fire. It was probably in competition with *H. erectus* that *Australopithecus* failed and became extinct.

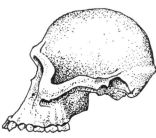

Reconstructed skeletons of hominids. From left to right *Australopithecus africanus* (height 1·3 m); *Australopithecus robustus* (height 1·5 m); *Homo sapiens* (height 1·7 m). (After Leakey).

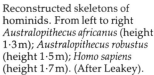

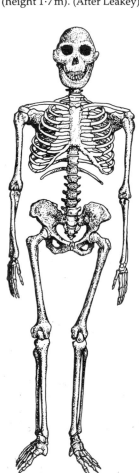

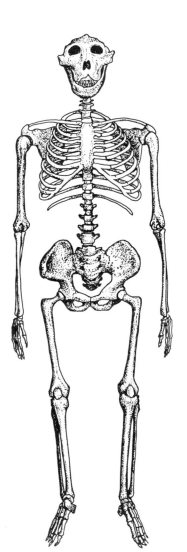

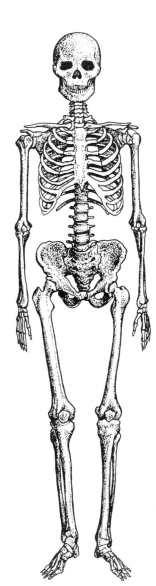

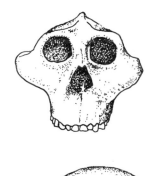

Skulls of *Australopithecus africanus* (above) and *Australopithecus robustus* (below), two early Pleistocene australopithecines. Their brain capacities are about 440 cm³ for *A. africanus* and 529 cm³ for *A. robustus*.

Restoration of *Homo erectus*.
Known as Java ape-man after the
first fossils discovered there.
Now known also from East Africa
and China, these hominids
manufactured Acheulean tools
and discovered the use of fire.

The final step, palaeontologically, in the emergence of man, was the appearance of our own species *H. sapiens*. His appearance in the mid Pleistocene has some overlap with *H. erectus*, as for example in the specimens of Heidelberg man. Two main subspecies or varieties of *H. sapiens* can be traced. The earlier Neanderthal are found around the Mediterranean littoral in abundance in late Pleistocene times, with other populations as far afield as Zambia and Indonesia. Neanderthal man was first found in the 1850s in Germany; he is shorter than modern man but with a rather larger brain; he had a prominently projecting jaw and a receding forehead – features sometimes seen in modern man. There is much debate as to whether Neanderthal and modern man could have interbred, but no agreed verdict. While Neanderthal man appears to have become extinct, there was some temporal and geographic overlap with modern man, at least in Europe.

The appearance of Upper Palaeolithic man, with his elaborate hunting ritual, advanced stone tools, and culture as expressed in cave art, marks the end of the fossil evidence. The emergence of the present-day geographic races of man are not readily traceable.

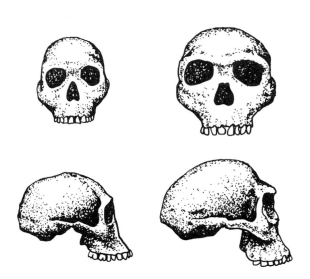

Skulls of *Homo habilis* (left) from the early Pleistocene of East Africa and *Homo erectus* (right) from the mid Pleistocene of Africa and Eurasia.

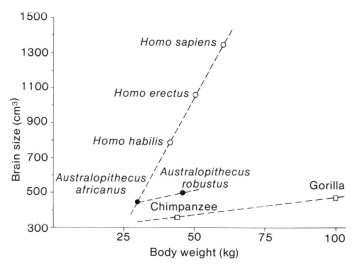

Graph to illustrate the relative increase in brain size to body size in modern man compared with his ancestors and with apes. (After Lewin).

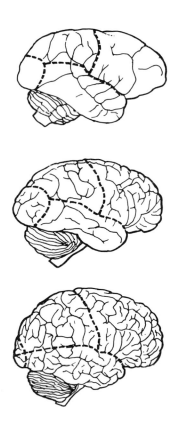

The brain of monkey (top), chimpanzee (middle) and modern man (above). Note the increase in the complexity of the surface features. All drawn to the same size.

Opposite: restoration of Neanderthal man, *Homo sapiens neanderthanensis*, the only known hominid inhabitant of the Mediterranean area and central Asia between 100 000 and 35 000 years ago. These short humans had brains with an average capacity of 1400 cm³ which slightly exceeds that of modern man. They buried their dead and hence their remains are plentifully fossilized.

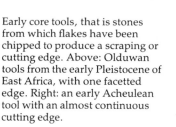

Early core tools, that is stones from which flakes have been chipped to produce a scraping or cutting edge. Above: Olduwan tools from the early Pleistocene of East Africa, with one facetted edge. Right: an early Acheulean tool with an almost continuous cutting edge.

	AUSTRALOPITHECUS robustus	*AUSTRALOPITHECUS africanus*	*HOMO habilis*	*HOMO erectus*	*HOMO sapiens*
Posture	Upright	Upright	Upright	Upright	Upright
Height m	1·3–1·6	1·0–1·3	1·2	1·6	1·75
Weight kg	35–55	20–40	35	50	70
Brain cm³ mean	500–530 520	425–490 440	590–775 675	750–1070 925	1000–2000 1360
Skull bones	Robust with sagittal crest; heavy brow ridges	Gracile without sagittal crest; brow ridges	Gracile without sagittal crest; brow ridges	Gracile without sagittal crest; slight brow ridges	Gracile without sagittal crest; no brow ridges
Dentition	Large molars and premolars with thick enamel	Large molars and premolars with thick enamel	Narrow molars, small premolars; thick enamel	Small molars and premolars	Small molars and premolars with thin enamel
Distribution	S + E Africa	S + E Africa	S + E Africa	Africa + Eurasia	Worldwide
Approx. time span MYA	2–1	3.5–2	2.5–1.5	1·5–0·5	0·5–0
Cultures	—	?Tools	Olduvan tools	Acheulean tools	Advanced tools

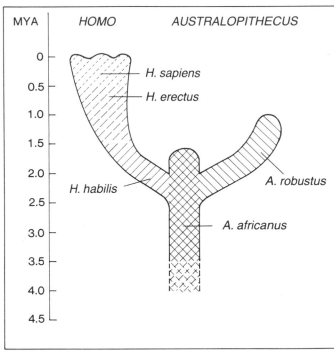

Family tree of the hominids.

Biologically we are an extraordinary species. In the space of about half a million years we have invaded and adapted to life in all parts of the planet; we have wrested control of almost all other forms of organic life – viruses still defeat us. We have a quite frightening breeding rate; humans being mass produced all over the world. We have evolved communications systems far beyond that of any other organism. We have harnassed the latent power of nature for almost unlimited energy resources. Yet we have failed to evolve social and political systems that would eliminate wars. Greed is probably our most deadly sin.

Most mammalian species have a life span of about a million years; few species alive today can be traced back far beyond our own origins. As a species we are already biologically middle aged. Will we aquire the wisdom to survive into old age?

The dinosauroid hominid; what might have evolved from dinosaurs had they inherited the earth. (After Russell & Séguin).

Glossary

Anaerobic Respiration in absence of free oxygen
Ankylosed Fused (of bones)
Apomorphic Derived or advanced character
Arboreal Tree dwelling
Astragalus Ankle bone
Auditory bulla Bone enclosing middle ear
Autapomorphic Uniquely derived character

Bilophodont Tooth with two transverse ridges
Brachiating Tree swinging
Brachydont Low crowned tooth
Brachyselenodont Tooth with low crown and crescentic cusps
Buccal Reference to outer or cheekward side of tooth
Bulla See Auditory bulla
Bunodont Tooth with domed cusps
Bunolophodont Low crowned tooth with domed cusps which are transversely linked by a loph or ridge
Bunoselenodont Tooth with inner cusps domed and outer cusps crescentic

Calcaneum Heel bone
Cannon bone Fused pair of foot bones
Carnassial Meat cutting tooth
Caudal Pertaining to the tail
Cementum Material of tooth roots; may also be present on tooth crown
Cerebellum Area of brain concerned with balance and muscular co-ordination
Cerebrum Anterior lobes or cerebral hemispheres of the brain; large in mammals and much folded
Cervical Of the neck
Cingulum An enamel ledge or rim around the base of the tooth crown
Cladistics A method of using primitive and derived characters to produce the branches or lineages of a phylogeny
Collagen Fibrous protein forming connective tissues
Conule Small cusp on crown of an upper tooth; equivalent on lower tooth is conulid
Coprolite Fossilized dung
Cursorial Running

Deciduous Of teeth, the milk set that are shed and replaced by the permanent teeth
Derived Of a character that is specialized or apomorphic
Diastema A gap between the anterior teeth and cheek teeth
Didactyl Two toed
Digitigrade Locomotion with only digits touching the ground
Diphyletic A stock made up of two lineages of separate origin; (compare with monophyletic)
Distal Distant from origin of structure; opposite of proximal

Echolocate Auditory perception using high or low frequency sound waves
Epicontinental Of seas, covering continental mass to edge of continental shelf

Epiphysis Ossified bone articulation, separated from shaft or main bone body (diaphysis) by cartilage during growth; the two fuse on maturity

Facet Smooth articular surface of a bone or occlusion surface of tooth
Flocculus Small lobe of the brain
Folivorous Leaf feeding
Foramen Opening or aperture
Fossa Cavity or depression on a bone
Fossorial Burrowing
Frugivore Fruit eating

Gnawers Animals such as rodents which persistently bite into hard substances with their front teeth
Gneiss Ancient crystalline metamorphic rock
Gondwana Former southern continent comprising South America, Africa, India, Australia and Antarctica
Graviportal Large heavily built mammals with massive limbs

Herbivore Animal feeding on vegetation
Holarctica Arctic regions of North America and Eurasia
Homodont Having all teeth in the jaw of similar pattern
Hypertrophy Increase in size
Hypselodont Continually growing open rooted high crowned cheek tooth
Hypsodont High crowned tooth; crown height greater than crown width

Igneous Primary rock originating from molten magma
Incisiform Tooth with the simple form of an incisor tooth
Innominate Pelvic bone formed by fusion of ilium, ischium and pubis

Labial As buccal, the outer or cheekward side of tooth
Lacustrine Of lakes
Laurasia Former northern continent comprising North America, Europe and Asia
Lingual Inner or tongue side of tooth
Loess Wind-blown dust
Loph Ridge joining two cusps on upper tooth; (lophid on lower tooth)

Meatus Short canal
Median On the midline
Metaloph Transverse loph linking metaconule and hypocone on posterior of upper molar
Metapodial Hand or foot bone; the former are metacarpals and the latter metatarsals
Molarize Of premolar tooth that acquires the form of a molar
Monocuspid Single cusped tooth
Monophyletic Stock of a single lineage all derived from a common ancestor

Neogene Unit comprising the Miocene and Pliocene divisions of the Tertiary

Neomorph New character

Neoteny Persistence of juvenile character in adult

Notochord Vertebrate skeletal rod; in mammals ossified as centra of the vertebrae

Nuchal Of the neck

Occlusion Surface contact of upper and lower teeth during mastication

Olfaction Sense of smell

Omnivore An eater of both animal and plant food

Orthal Up and down jaw movement

Pachyostosis Thick dense bone

Palaeogene Unit comprising the Palaeocene, Eocene and Oligocene divisions of the Tertiary

Palatal Of the palate or roof of the mouth

Pangaea Former supercontinent comprising all land masses

Parastyle Small tooth cusp anterior to paracone

Parturition Birth

Pelagic Organisms which drift or swim in water; in contrast to benthic organisms living on the bottom at the water/sediment interface

Pentadactyl Limb with five digits

Permafrost Permanently frozen ground

Petroglyph Incised rock art

Petrosum Bone of inner ear

Phororhachids Family of extinct carnivorous South American birds

Phylogeny Evolutionary history

Plantigrade Locomotion with metapodials (palms and soles) of feet touching the ground

Plesiomorphic Primitive character

Polydont With many teeth

Postorbital Bone behind the eye

Procumbent Of incisor tooth which are inclined forward

Propalinal Fore and aft jaw movement

Proximal Close to origin of structure

Rostral bones Bones forming roof of nasal cavity

Rugosity Wrinkled

Sagittal Fore and aft section in the mid-line

Scansorial Climbing

Schist Ancient metamorphic rock with parallel alignment of minerals

Sectorial Of tooth, adapted to cutting

Selenodont Of tooth with crescentic cusps

Sensu lato In the wide sense

Spatulate Of incisor tooth which are spade-shaped

Sternum Breast plate

Style Small upper tooth cusp; stylid on lower tooth

Subhypsodont Tooth crown height intermediate between brachydont and hypsodont

Talonid Heel or posterior part of the crown of lower molar tooth

Taxon (Plural *Taxa*) A taxonomic group of any rank

Tethys Ancient Mediterranean Sea

Trenchant Of tooth with cutting features

Ungulate Hoofed mammal

Unguligrade Locomotion with only the hoof touching the ground

Vascularized Richly supplied with blood vessels

Vibrissa (Plural *Vibrissae*) Whiskers or elongate hairs on the upper lip

Volcanoclastic Sediment composed of volcanic ejecta

Zygodont Of proboscidean tooth with high pitched conical cusps

Zygolophodont Of proboscidean tooth with high pitched traverse lophs or ridges

Zygoma Cheek bone

Further reading

General and Chapters 1–3

CLUTTON-BROCK, J. 1981. *Domesticated animals from early times.* 208 pp. Heinemann & British Museum (Nat. Hist.), London.

COLBERT, E. H. 1961. *Evolution of the vertebrates.* 479 pp. Wiley & Sons, New York.

CORBET, G. B. & HILL, J. E. 1980. *A world list of mammalian species.* 226 pp. British Museum (Nat. Hist.), London.

GREENWOOD, P. H. (Editor). 1981. *The evolving earth* (280 pp.) and *The evolving biosphere* (320 pp.) 2 vols. British Museum (Nat. Hist.), London and Cambridge University Press, Cambridge.

KURTEN, B. 1968. *Pleistocene mammals of Europe.* 317 pp. Weidenfeld & Nicolson, London.

KURTEN, B. 1971. *The age of mammals.* 250 pp. Weidenfeld & Nicolson, London.

KURTEN, B. & ANDERSON, E. 1980. *Pleistocene mammals of North America.* 442 pp. Columbia University Press, New York.

MCKENNA, M. C. 1975. Toward a phylogenetic classification of the Mammalia. pp. 21–46, *in* Luckett, W. P. & Szalay, F. S. (eds) *Phylogeny of the primates.* Plenum Press, New York.

MAGLIO, V. J. & COOKE, H. B. S. 1978. *Evolution of African mammals.* 641 pp. Harvard Univ. Press, Cambridge and London.

PATTERSON, C. 1978. *Evolution.* 197 pp. British Museum (Nat. Hist.), London.

PIVETEAU, J. (Editor) 1958 and 1961. *Traité de Paléontologie: Mammifères.* Tome VI, 2 vols. Vol. 1 (1961) 1138 pp., Vol. 2 (1958) 962 pp.

ROMER, A. S. 1966. *Vertebrate paleontology* (Third edition). 468 pp. Chicago Univ. Press, Chicago and London.

SAVAGE, D. E. & RUSSELL, D. E. 1983. *Mammalian paleofaunas of the world.* 432 pp. Addison-Wesley Publ. Co., Reading, Mass. and London.

SCOTT, W. B. 1937. *A history of land mammals in the western hemisphere.* 786 pp. Macmillan, New York.

SIMPSON, G. G. 1945. The principles of classification and a classification of mammals. *Bull. Am. Mus. nat. Hist.* **85:** 1–350.

THENIUS, E. 1969. *Phylogenie der Mammalia; Stammesgeschichte der Säugetiere (einschliesslich der Hominiden).* 722 pp. Walter de Gruyer & Co., Berlin.

WALKER, E. P. 1975. *Mammals of the world* (Third edition). 2 vols. 1500 pp. Johns Hopkins Press, Baltimore.

ZEUNER, F. E. 1963. *A history of domesticated animals.* 560 pp. Hutchinson, London.

Chapter 4   Reptiles into mammals

KEMP, T. S. 1982. *Mammal-like reptiles and the origin of mammals.* 363 pp. Academic Press, London.

KERMACK, D. M. & K. A. 1984. *The evolution of mammalian characters.* 149 pp. Croom Helm, London.

LILLEGRAVEN, J. A., KIELAN-JAWOROWSKA, Z. & CLEMENS, W. A. 1979. *Mesozoic mammals; the first two-thirds of mammalian history.* 311 pp. Univ. Calif. Press, Berkeley.

Chapter 5   Insectivores

BUTLER, P. M. 1981. The giant erinaceid insectivore *Deinogalerix* Freudenthal, from the Upper Miocene of Gargano, Italy. *Scripta Geol.* **57:** 1–72.

EMRY, R. J. 1970. A North American Oligocene pangolin and other additions to the Pholidota. *Bull. Am. Mus. nat. Hist.* **142:** 455–510.

KOENIGSWALD, W. von 1980. Das Skelett eines Pantolestiden (Proteutheria, Mamm.) aus mittleren Eozän von Messel bei Darmstadt. *Paläont. Z.,* **54:** 268–87.

LUCKETT, W. P. (Editor) 1980. *Comparative biology and evolutionary relationships of tree shrews.* 314 pp. Plenum Press, New York and London.

PATTERSON, B. 1965. The fossil elephant shrews (Family Macroscelidae). *Bull. Mus. comp. Zoo. Harvard* **133:** 295–335.

PATTERSON, B. 1975. The fossil aardvarks (Mammalia: Tubulidentata). *Bull. Mus. comp. Zoo. Harvard* **147:** 185–237.

PICKFORD, M. 1975. New fossil Orycteropodidae (Mammalia, Tubulidentata) from East Africa. *Neth. J. Zool.* **25:** 57–88.

REED, K. M. 1961. The Proscalopinae, a new subfamily of talpid insectivores. *Bull. Mus. comp. Zool. Harvard* **125:** 473–94.

RUSSELL, D. E., LOUIS, P. & SAVAGE, D.E. 1975. Les Adapisoricidae de l'Éocène inférieur de France. *Bull. Mus. natl. Hist. nat., Sci. Terre,* No. 45 (327): 129–94.

SCHMIDT-KITTLER, N. 1973. Dimyloides-Neufunde aus der oberoligozänen Spaltenfüllung 'Ehrenstein 4' (Süddeutschland). *Mitt. Bayer. St. Paläont. hist. Geol.* **13:** 115–39.

STORCH, G. 1978. *Eomanis waldi,* ein Schuppentier aus dem Mittel-Eozaen der 'Grube Messel' bei Darmstadt (Mammalia, Pholidota). *Senckenberg. leth.* **59:** 503–29.

STORCH, G. 1981. *Eurotamandua joresi,* ein Myrmecophagide aus der Eozän der 'Grube Messel' bei Darmstadt (Mammalia, Xenarthra). *Senckenberg. leth.* **61:** 247–89.

SZALAY, F. S. 1968. Origins of the Apatemyidae (Mammalia, Insectivora). *Am. Mus. Novit.* **2352:** 1–11.

SZALAY, F. S. 1969. Mixodectidae, Microsyopidae and the insectivore-primate transition. *Bull. Am. Mus. nat. Hist.* **140:** 193–330.

VAN VALEN, L. 1966. Deltatheridia, a new order of mammals. *Bull. Am. Mus. nat. Hist.* **132:** 1–126.

VAN VALEN, L. 1967. New Paleocene insectivores and insectivore classification. *Bull. Am. Mus. nat. Hist.* **135:** 217–84.

Chapter 6   Carnivores

ARCHER, M. (Editor) 1982. *Carnivorous marsupials.* Vol. 1, 396 pp., Vol. 2, 407 pp. Roy. Soc. New South Wales, Sydney.

BUGGE, J. 1978. The cephalic arterial system in carnivores. *Acta anat.* **101:** 45–61.

EMERSON, S. B. & RADINSKY, L. 1980. Functional analysis of sabretooth cranial morphology. *Paleobiol.* **6:** 295–312.

GINSBURG, L. 1980. *Hyainailoros sulzeri,* mammifère créodonte du Miocène d'Europe. *Annls. Paléont. (Vert.)* **66:** 19–73.

MARSHALL, L. G. 1976. Evolution of the Thylacosmilidae, fossil marsupial 'sabretooths' of South America. *Paleobios* **23:** 1–31.

MARSHALL, L. G. 1978. Evolution of the Borhyaenidae, extinct South American predaceous marsupials. *Univ. Calif. Publ. geol. Sci.* **117:** 1–89.

MELLET, J. S. 1977. Paleobiology of North American *Hyaenodon* (Mammalia, Creodont). *Contr. vertebr. Evol.* **1:** 134 pp.

RADINSKY, L. 1977. Brains of early carnivores. *Paleobiol.* **3:** 333–49.

SAVAGE, R. J. G. 1977. Evolution in carnivorous mammals. *Palaeontology* **20**: 237–71.

SPRINGHORN, R. 1980. *Paroodectes feisti*, der erste Miacide (Carnivora, Mammalia) aus dem Mittel-Eozän von Messel. *Palaont. Z.* **54**: 171–98.

SPRINGHORN, R. 1982. Neue Raubtiere (Mammalia: Creodonta et Carnivora) aus dem Lutetium der Grube Messel (Deutschland). *Palaeontographica A* **179**: 105–41.

Chapter 7   Paddlers and swimmers

General

Symposium on advances in systematics of marine mammals. 1976. *Syst. Zool.* **25**, part 4: 301–436. [Ten papers, multiauthored].

ROBINSON, J. A. 1975. The locomotion of Plesiosaurs. *N. Jb. Geol. Palaont. Abh.* **149**: 286–332.

Sirenians and desmostylians

DOMNING, D. P. 1978. Sirenian evolution in the North Pacific Ocean. *Univ. Calif. Publ. geol. Sci.* **118**: 1–176.

DOMNING, D. P., MORGAN, G. S. & RAY, C. E. 1982. North American Eocene sea cows (Mammalia: Sirenia). *Smithson. Contrib. Paleobiol.* **52**: 1–69.

MCCOY, E. E. & HECK, K. L. 1976. Biogeography of corals, seagrasses and mangroves. *Syst. Zool.* **25**: 201–10.

REINHART, R. H. 1959. A review of the Sirenia and Desmostylia. *Univ. Calif. Publ. geol. Sci.* **36**: 1–146.

REINHART, R. H. 1976. Fossil sirenians and desmostylids from Florida and elsewhere. *Bull. Florida St. Mus. biol. Sci.* **20**: 187–300.

SHIKAMA, T. 1966. Postcranial skeletons of Japanese Desmostylia. *Spec. Pap. palaeont. Soc. Japan* **12**: 1–202.

SIMPSON, G. G. 1932. Fossil Sirenia of Florida and the evolution of the Sirenia. *Bull. Am. Mus. nat. Hist.* **59**: 419–503.

Pinnipeds

BARNES, L. G. 1972. Miocene Desmatophocinae (Mammalia: Carnivora) from California. *Univ. Calif. Publ. geol. Sci.* **89**: 1–68

KING, J. E. *Seals of the world* Second edition. 240 pp. British Museum (Nat. Hist.), London and Oxford University Press, Oxford.

MITCHELL, E. & TEDFORD, R. H. 1973. The Enaliarctinae a new group of extinct aquatic Carnivora and a consideration of the origin of the Otariidae. *Bull. Am. Mus. nat. Hist.* **151**: 201–85.

REPENNING, C. A., RAY, C. E. & GRIGORESCU, D. 1979. Pinniped biogeography. pp. 357–69 *in* Gray, J. & Boucot, A. J. (Eds.) *Historical biogeography, plate tectonics and the changing environment*. Oregon State Univ. Press.

REPENNING, C. A. & TEDFORD, R. H. 1977. Otarioid seals of the Neogene. *U.S. geol. Surv. Prof. Pap.* **992**: 1–93.

SARICH, V. M. 1969. Pinniped phylogeny. *Syst. Zool.* **18**: 416–22.

Cetaceans

FORDYCE, R. E. 1980. Whale evolution and Oligocene Southern Ocean environments. *Palaeogeogr. Palaeoclimat. Palaeoecol.* **31**: 319–36.

GASKIN, D. E. 1976. The evolution, zoogeography and ecology of Cetacea. *Oceanogr. Mar. Biol. Ann. Rev.* **14**: 247–346.

GINGERICH, P. D. *et al.* 1983. Origin of whales in epicontinental remnant seas: new evidence from the early Eocene of Pakistan. *Science, N.Y.* **220**: 403–06.

KELLOGG, R. 1936. A review of the Archaeoceti. *Publ. Carnegie Instn* **482**: 1–366.

MCHEDLIDZE, G. A. 1975. A review of the historical development of the Cetacea. *J. palaeont. Soc. India.* **20**: 81–88.

MEAD, J. G. 1975. A fossil beaked whale (Cetacea: Ziphidae) from the Miocene of Kenya. *J. Paleont.* **49**: 745–51.

NESS, A. R. 1967. A measure of asymmetry of the skulls of odontocete whales. *J. Zool. Lond.* **153**: 209–21.

ROTHAUSEN, K. 1967. Die Klimabindung der Squalodontoidea (Odontoceti, Mamm.) und anderer mariner Vertebrata. *Sonderveroff. Geol. Inst. Univ. Koln* **13**: 157–66.

Chapter 8   Gliders and fliers

DAL PIAZ, G. 1937. I mammiferi dell'Oligocene Veneto. *Archaeopteropus transiens. Mem. Ist. Geol. Univ. Padova,* **11; 6**: 1–8.

HILL, J. E. & SMITH, J. D. 1984. *Bats a natural history.* 192 pp. British Museum (Nat. Hist.), London.

JEPSEN, G. L. 1970. Bat origins and evolution. 64 pp. *in* Wimsatt, W. A. (Ed.) *Biology of bats* Vol. 1. Academic Press, New York & London.

PENNYCUICK, C. J. 1972. *Animal flight.* Studies in Biology No. 33. 68 pp. Edward Arnold, London.

ROSE, K. D. & SIMONS, E. L. 1977. Dental function in the Plagiomenidae: origin and relationships of the mammalian order Dermoptera. *Contr. Mus. Paleont. Univ. Mich.* **24**: 221–36.

SMITH, J. D. & STORCH, G. 1981. New middle Eocene bats from 'Grube Messel' near Darmstadt, W-Germany. *Senckenberg. biol.* **61**: 153–67.

Chapter 9   Gnawers

CHALINE, J. 1974. Esquisse de l'évolution morphologique, biométrique et chromosomique du genre *Microtus* (Arvicolidae, Rodentia) dans le Pléistocene de l'hemisphere nord. *Bull. soc. géol. Fr.* **16**: 440–50.

DAWSON, M. R. 1967. Lagomorph history and the stratigraphic record. *Univ. Kansas Dept. Geol. Spec. Publ.* **2**: 30 pp.

MARTIN, L. D. & BENNETT, D. K. 1977. The burrows of the Miocene beaver *Palaeocastor*, western Nebraska, USA. *Palaeogeogr. Palaeoclimat. Palaeoecol.* **22**: 173–93.

SHOTWELL, J. A. 1958. Evolution and biogeography of the aplodontid and mylagaulid rodents. *Evolution* **12**: 451–84.

STEHLIN, H. G. & SCHAUB, S. 1951. Die Trigonodontie der simplicidentaten Nager. *Schweiz. paläont. Abh.* **67**: 1–385.

STIRTON, R. A. 1935. A review of the Tertiary beavers. *Univ. Calif. Publ. geol. Sci.* **23**: 391–458.

TOBIEN, H. 1978. Brachyodonty and hypsodonty in some Paleogene Eurasian lagomorphs. *Mainzer. geowiss. Mitt.* **6**: 161–75.

WOOD, A. E. 1962. The early Tertiary rodents of the family Paramyidae. *Trans. Am. phil. Soc.* **52**: 1–261.

WOOD, A. E. 1974. The evolution of the Old World and New World hystricomorphs. *Symposia zool. Soc. Lond.* **34**: 21–60.

Chapter 10   Early rooters and browsers

Arctocyonids

KOENIGSWALD, W. von 1983. Skelettfunde von *Kopidodon* (Condylarthra, Mammalia) aus dem mitteleozänen Ölschiefer von Messel bei Darmstadt. *N. Jb. Geol. Palaont., Abh.* **167**: 1–39.

Mesonychids

LUCAS, S. G. 1982. The phylogeny and composition of the order Pantodonta (Mammalia, Eutheria). *Proc. Third N. Am. Paleont. Convention,* Vol. **2**: 337–42.

SIMONS, E. L. 1960. The Paleocene Pantodonta. *Trans. Am. phil Soc.* **50**: 1–81.

SZALAY, F. S. 1969. Origin and evolution of function of the mesonychid feeding mechanism. *Evolution* **23**: 703–20.

SZALAY, F. S. & GOULD, S. J. 1966. Asiatic Mesonychidae (Mammalia, Condylarthra). *Bull. Am. Mus. nat. Hist.* **132**: 127–74.

Uintatheres

LUCAS, S. G. 1982. A review of Chinese uintatheres and the origin of Dinocerata (Mammalia, Eutheria). *Proc. Third N. Am. Paleont. Convention,* Vol. **2**: 551–56.

MARSH, O. C. 1885. Dinocerata, a monograph on an extinct order of gigantic mammals. *U.S. Geol. Surv. Monog.* **10**: 1–237.

WHEELER, W. H. 1961. Revision of the uintatheres. *Bull. Peabody Mus. nat. Hist.* **14**: 1–93.

Taeniodonts

PATTERSON, B. 1949. Rates of evolution in Taeniodonts. pp. 243–78 *in:* Jepsen, G. L., Simpson, G. G. & Mayer, E. *Genetics, paleontology and evolution.* Princeton Univ. Press.

SCHOCH, R. M. 1981. Taxonomy and biostratigraphy of the early Tertiary Taeniodonta (Mammalia: Eutheria). *Bull. geol. Soc. Am.* **92**: Pt 2: 1982–2267.

Tillodonts

GAZIN, C. L. 1953. The Tillodontia: an early Tertiary order of mammals. *Smithson. Misc. Collns.,* **121**, No. 10, 1–110.

GINGERICH, P. D. & GUNNELL, G. F. 1979. Systematics and evolution of the genus *Esthonyx* (Mammalia, Tillodontia) in the early Eocene of North America. *Contr. Mus. Paleont. Univ. Mich.* **25**: 125–53.

Condylarths

GAZIN, C. L. 1965. A study of the early Tertiary condylarthran mammal *Meniscotherium. Smithson. Misc. Collns.* **149**, No. 2: 1–98.

GAZIN, C. L. 1968. A study of the Eocene condylarthran mammal *Hyopsodus. Smithson. Misc. Collns.* **153**, No. 4: 1–89.

Embrithopods

ANDREWS, C. W. 1906. *A descriptive catalogue of the Tertiary Vertebrata of the Fayum, Egypt.* 324pp. British Museum (Nat. Hist.), London.

SEN, S. & HEINTZ, E. 1979. *Palaeoamasia kansui* Ozansoy 1966, Embrithopode (Mammalia) de l'Éocène d'Anatolie. *Annls. Paleont., (Vert.)* **65**: 73–91.

Proboscideans

OSBORN, H. F. 1936 & 1942. *Proboscidea,* 2 vols. 1676 pp. Am. Mus. nat. Hist. Press, New York.

TASSY, P. 1981. Le crane de *Moeritherium* (Proboscidea, Mammalia) de l'Éocène de Dor el Talha (Libya). *Bull. Mus. natl. Hist. nat.,* Paris **3C**: 87–147.

TASSY, P. 1982. Les principales dichotomies dans l'histoire des Proboscidea (Mammalia): une approche phylogénétique. *Geobios mem. Spéc.* **6**: 225–45.

TASSY, P. 1983. Les Elephantoidea Miocènes du plateau du Potwar, Groupe de Siwalik, Pakistan. *Annls. Paléont. (Vert.)* **69**: 99–136, 235–97, 317–54.

WELLS, N. A. & GINGERICH, P. D. 1983. Review of Eocene Anthracobundiae (Mammalia, Proboscidea). *Contr. Mus. Paleont. Univ. Mich.* **26**: 117–39.

Chapter 11   Mammals on island continents

ARCHER, M. & CLAYTON, G. (Editors) 1984. *Vertebrate zoogeography and evolution in Australia (Animals in space and time).* 1203 pp. Hesperian Press, Carlisle, W. Australia.

COOMBS, M. C. 1983. Large mammalian clawed herbivores: a comparative study. *Trans. Am. phil. Soc.* **73**, pt 7: 1–96.

LOOMIS, F. B. 1914. *The Deseado Formation of Patagonia.* 232 pp. Amherst College Press, Amherst, Mass.

PATTERSON, B. & PASCUAL, R. 1968. Evolution of mammals on southern continents. V. The fossil mammal fauna of South America. *Quart. Rev. Biol.* **43**: 409–51.

RICH, P. V. & THOMPSON, E. M. (Editor) 1982. *The fossil vertebrate record of Australia.* 759 pp. Monash Univ. Press, Clayton, Australia.

SIMPSON, G. G. 1948. The beginning of the age of mammals in South America. Part 1. *Bull. Am. Mus. nat. Hist.* **91**: 1–232.

SIMPSON, G. G. 1967. The beginning of the age of mammals in South America. Part 2. *Bull. Am. Mus. nat. Hist.* **137**: 1–260.

SIMPSON, G. G. 1980. *Splendid isolation; the curious history of South American mammals.* 266 pp. Yale Univ. Press, New Haven.

Chapter 12   Hoofed herbivores

Perissodactyls

MACFADDEN, B. J. 1976. Cladistic analysis of primitive equids, with notes on other perissodactyls. *Syst. Zool.* **25**: 1–14.

OSBORN, H. F. 1929. The Titanotheres of ancient Wyoming, Dakota and Nebraska. *U.S. geol. Surv. Monograph* **55**: 2 vols, 953 pp.

RADINSKY, L. 1963. Origin and early evolution of North American Tapiroidea. *Bull. Peabody Mus. nat. Hist.* **17**: 1–106.

RADINSKY, L. 1965. Early Tertiary Tapiroidea of Asia. *Bull. Am. Mus. nat. Hist.* **129**: 181–264.

RADINSKY, L. 1967. A review of the rhinocerotoid family Hydracodontidae (Perissodactyla). *Bull. Am. Mus. nat. Hist.* **136**: 1–46.

RADINSKY, L. 1968. The early evolution of the Perissodactyla. *Evolution* **23**: 308–28.

SIMPSON, G. G. 1951. *Horses.* 247 pp. Oxford Univ. Press, Oxford.

ZAPFE, H. 1979. *Chalicotherium grande* (Blainv.) aus der miozaenen Spaltenfuellung von Neudorf an der Maech, Tschechoslowakei. *Neue Denkschr. Naturhist. Mus. Wien* **2**: 282 pp.

Artiodactyls

CHAPMAN, D. I. 1975. Antlers – bones of contention. *Mammal Rev.,* **5**: 121–72.

COLBERT, E. H. 1935. Siwalik Mammals in the American Museum of Natural History. *Trans. Am. phil. Soc.* **26**: 1–401.

FRICK, C. 1937. Horned ruminants of North America. *Bull. Am. Mus. nat. Hist.* **69**: 1–669.

FRICK, C. & TAYLOR, B. E. 1968. A generic review of the stenomyline camels. *Am. Mus. Novit.* **2353**: 1–51.

GAZIN, C. L. 1955. A review of the upper Eocene Artiodactyla of North America. *Smithson. Miscell. Collns.* **128**, No. 8: 1–96.

HAMILTON, W. R. 1978. Fossil giraffes from the Miocene of Africa and a revision of the phylogeny of the Giraffoidea. *Phil. Trans. Roy. Soc. Lond.* **283B**: 165–229.

HARRIS, J. M. & WHITE, T. D. 1979. Evolution of the Plio-Pleistocene African Suidae. *Trans. Am. phil. Soc.* **69**, Part 2: 1–128.

JANIS, C. 1982. Evolution of horns in ungulates: ecology and paleoecology. *Biol. Rev.* **57**: 261–318.

PATTON, T. H. & TAYLOR, B. E. 1971. The Synthetoceratinae (Mammalia, Tylopoda, Protceratidae). *Bull. Am. Mus. nat. Hist.* **145**: 119–218.

PATTON, T. H. & TAYLOR, B. E. 1973. The Protoceratinae (Mammalia, Tylopoda, Protoceratidae) and the systematics of the Protoceratidae. *Bull. Am. Mus. nat. Hist.* **150**: 347–414.

ROGERS, T. A. 1958. The metabolism of ruminants. *Scient. Am.* **198**: 34–38.

SUDRE, J., RUSSELL, D. E., LOUIS, P. & SAVAGE, D. E. 1983. Les Artiodactyles de l'Éocène inférieur d'Europe. *Bull. Mus. natl. Hist. nat Paris* **5**: 281–333 and 339–365.

TAYLOR, B. E. & WEBB, S. D. 1976. Miocene Leptomerycidae (Artiodactyla, Ruminantia) and their relationships. *Am. Mus. Novit.* **2596**: 1–22.

THENIUS, E. 1970. Zur Evolution und Verbreitungsgeschichte der Suidae (Artiodactyla, Mammalia). *Z. Säugetierkunde* **35**: 321–42.

THEWISSEN, J. G. M., RUSSELL, D. E., GINGERICH, P. D. & HAUSSAIN, S. T. 1983. A new dichobunid artiodactyl (Mammalia) from the Eocene of North-west Pakistan. *Proc. K. ned. Akad. Wet.* **B 86**: 153–80.

WEBB, S. D. 1965. The osteology of *Camelops. Bull. Los Ang. Cty. Mus. Sci.* **1**: 1–54.

WEBB, S. D. 1972. Locomotor evolution in camels. *Forma funct.* **5**: 99–112.

WEBB, S. D. & TAYLOR, B. E. 1980. The phylogeny of hornless ruminants and a description of the cranium of *Archaeomeryx. Bull. Am. Mus. nat. Hist.* **167**: 117–58.

Chapter 13   Of men and monkeys

ANON. 1980. *Man's place in evolution.* 108 pp. British Museum (Nat. Hist.), London and Cambridge University Press, Cambridge.

ISAAC, G. & LEAKEY, R. E. F. (Editors) 1979. *Human ancestors.* Readings from Scientific American. 130 pp. Freeman & Co. San Francisco.

JERISON, H. J. 1973. *Evolution of the brain and intelligence.* 482 pp. Academic Press, New York & London.

LEWIN, R. 1984. *Human evolution; an illustrated introduction.* 104 pp. Blackwells Scient. Publ., Oxford.

NAPIER, J. R. & P. H. 1985. *The natural history of the primates.* 200 pp. British Museum (Nat. Hist.), London.

SZALAY, F. S. & DELSON, E. 1979. *Evolutionary history of the primates.* 580 pp. Academic Press, New York & London.

TATTERSALL, I. 1970. *Man's ancestors: an introduction to primate and human evolution.* 64 pp. John Murray, London.

Index

THE AGE OF MAMMALS

PERIODS	AGE MYA	LAND MAMMAL AGES NORTH AMERICAN / EUROPEAN	
HOLOCENE		RANCHOLABREAN	OLDENBURGIAN
PLEISTOCENE	2.5	IRVINGTONIAN	BIHARIAN
PLIOCENE	5	BLANCAN	VILLAFRANCHIAN
			RUSCINIAN
MIOCENE		HEMPHILLIAN	TUROLIAN
		CLARENDONIAN	VALLESIAN
		BARSTOVIAN	ASTARACIAN
		HEMINGFORDIAN	ORLEANIAN
			AGENIAN
	24	ARIKAREEAN	
OLIGOCENE		GERINGIAN	ARVERNIAN
	37	WHITNEYAN ORELLAN CHADRONIAN DUCHESNIAN	SUEVIAN
			HEADONIAN
EOCENE		UINTAN	
		BRIDGERIAN	LUTETIAN
	54	WASATCHIAN	CUISIAN
		CLARKFORKIAN	CERNAYSIAN
PALAEOCENE		TIFFANIAN	
		TORREJONIAN	
	65	PUERCAN	